Choke Hold:
The Fossil Fuel Industry's Fight Against Climate Policy, Science and Clean Energy

By Neela Banerjee

David Hasemyer

Marianne Lavelle

Robert McClure

and Brad Wieners

Edited by Clark Hoyt

TABLE OF CONTENTS

INTRODUCTION

L.J. Turner is a rancher in Wyoming who lost the freshwater his grandfather bequeathed him to the strip mines of the big coal companies in the Powder River Basin. Bryan Latkanich can't drink his well water anymore, and he is sure the fracking rigs he allowed on his property in rural Pennsylvania are to blame. Diane Eckhardt, a peach farmer in central Texas, has watched her crop fail in the warming climate, while her congressman denies the problem even exists.

And there's Bethel Brock, a coal miner from Virginia, who had to fight against company doctors and lawyers for 14 years to get the black lung benefits he was due under federal law. He needed the help of our reporting to secure them.

L.J., Bryan, Diane and Bethel don't know each other, but they have all had their lives upended by the same force. They help us tell a series of large and complicated stories in simple, human terms.

The stories in this book expose the depth and breadth of the fossil fuel industry's grip over these individuals—and over the richest society in the world and, by extension, the whole globe. They explain how the industry guards its interests at every turn and, most consequentially, how it has marshaled its inordinate wealth and influence to stop action that would curtail the burning of fossil fuels, the source of its prosperity and the main cause of climate change.

InsideClimate News embarked on this project in January of 2017 when the fossil fuel industry gained an unrestrained ally in the Oval Office, unafraid to give credence to extreme positions on climate science once rejected as too far-fetched. We set out to understand how the industry has been able to succeed for so long and on such a grand scale despite accumulating planetary costs and consequences.

Calling our series "Choke Hold" because we found the industry's grip to be so tight, we compiled a chronicle of its fight against climate policy, science and clean energy—and how it suffocates the aspirations of ordinary Americans who happen to be in the way.

L.J., Bryan, Diane and Bethel helped us bring life to a project that otherwise might have constituted a dry treatise. We combined their human narratives with explanation, investigation and graphic art, and then made our storytelling richer by featuring each of them in a short documentary video. We shared our work for free in publishing partnerships with media outlets across the country to reach as broad an audience as possible.

Industry applies its choke hold mostly in plain view. It hires expensive law firms, controls regulatory policy and enforcement, and finances the campaigns of elected officials. It maintains its grip thanks to long-standing tax breaks and lucrative subsidies, by undermining the progress of alternative energy, and through persistent and ubiquitous misinformation.

Our stories tell about all these policies and practices. Each one is a case study of its own, an examination of the choke hold, one finger at a time.

We believe this is a timely and evergreen body of work. The choke hold operates with collective force and efficiency in ways still invisible to most people, at a time when it needs to be most seen.

* You'll need to go to
https://insideclimatenews.org/content/choke-hold to see
the full multi-media package.

WITH BARE KNUCKLES AND BIG DOLLARS, EXXON FIGHTS CLIMATE PROBE TO A LEGAL STALEMATE

A shift in venue and lost emails now present new challenges for the company's lawyers who buy time and protect profits by playing defense with offense.

By David Hasemyer
June 5, 2017

Ted Wells, one of the nation's most prominent litigators for big corporations, was about to win again as he sat with his team in a Dallas courtroom last fall, representing ExxonMobil. U.S. District Judge Ed Kinkeade looked their way and joked, "Y'all have 300 lawyers on your side."

Wells, 66, had come before Kinkeade to thwart fraud investigations launched by the attorneys general of New York and Massachusetts, who are looking into whether the mammoth oil company has misled investors and the public for years about the dangers of climate change.

Kinkeade, with his folksy joshing and pointed comments, made little secret of his sympathies. He kidded that his horse was tied up outside and he might need an interpreter to pierce the Boston accent of the Massachusetts counsel. He wondered aloud if those Northern officials would be as worried about the climate if their states had as much oil as his native Texas. "I'm just saying, think about it."

A little more than three weeks later, he handed Exxon a major victory, ruling that Massachusetts Attorney General Maura Healey may have acted in "bad faith." Some legal scholars said the order could give Wells unprecedented license to try to prove that Healey was part of a conspiracy to silence Exxon in the debate over climate policy.

The investigators had suddenly become the investigated, a striking example of a legal strategy of massive resistance to any effort to hold Exxon accountable for global warming or harm to the environment. The message from Exxon, delivered by Wells and his army of lawyers: Mess with us at your peril.

Exxon is not alone. Its defiant stance is typical of the fossil fuel industry's fight against stronger government oversight and efforts to protect public health. Fossil fuel corporations doggedly battle lawsuits, regulations and investigations, because it allows them to continue pumping oil, mining coal and fracturing for natural gas—and banking the profits— even if sometime later there will be a reckoning, said Christine Todd Whitman, former administrator of the Environmental Protection Agency under President George W. Bush.

"It's a decision they make based on hard numbers," she said in an interview.

Wells pursued a similar course years ago, when he represented Philip Morris against government charges that it hid the health dangers of smoking. The tobacco industry eventually lost and paid dearly, but its long war of attrition bought time.

The Exxon case took an unexpected detour in late March, when Kinkeade, perhaps recognizing that he was on thin ice because the alleged conspiracy hadn't taken place within his jurisdiction, transferred the case to federal court in New York. The company starts over with a new judge, Valerie

Caproni, who signaled she is less sympathetic to its arguments than Kinkeade was.

Still, there will be more filings and motions and more costly delay for Healey and a coalition of state attorneys general who announced more than a year ago their intention to take on the fossil fuel industry over its role in climate change.

Wells and Exxon declined to comment for this article.

A review of their fight against the attorneys general shows how tenacious—some opponents say ruthless—Exxon can be:

- The tiny Virgin Islands dove into an investigation alongside Healey and New York Attorney General Eric Schneiderman and was quickly driven to surrender when the company and an allied conservative think tank filed lawsuits against its attorney general. They charged abridgement of their constitutional rights. After that, no other attorney general started new investigative action, and the attorney general of Maryland says their coalition now exists "in name only."

- Exxon's subpoenas went far beyond the attorneys general and demanded information from non-profit environmental groups and even private individuals active in the national climate debate. Those targeted saw the subpoenas as an effort to intimidate them. A member of an Exxon advisory board resigned, saying that the company was not acting as a good corporate citizen by pressuring those organizations.

- The revelation that former Exxon chief executive Rex Tillerson used a separate email account, with the alias

"Wayne Tracker," to discuss the implications of climate change with his board of directors and key executives could put the company back on the defensive. But, even as Wells acknowledged that some of the Wayne Tracker e-mails may not have been preserved, he counter-attacked in familiar fashion. He declared that New York's demand for those emails was part of "a highly politicized and bad faith investigation."

This article is the first of a series examining the formidable power the fossil fuel industry deploys across society to protect its interests. It fights adversaries as small as a family in Louisiana suing over a natural gas pipeline that polluted their property and as large as the federal government, with its rules aimed at reducing carbon emissions nationwide. The industry sowed doubt for decades about climate science, spending $2.9 billion on advocacy advertising alone in a 10-year period ending in 2015. It spent $1.3 billion more lobbying to shape public policy on energy issues during the same period and has pumped out $827.9 million in campaign contributions since 2000 to elect sympathetic officials at the local, state and federal levels.

In an ironic twist, industry backers, who typically declare devotion to free markets, have sought to weaken or even ban competition from other sources of energy. A bill authored by coal supporters in Wyoming would have forbidden electricity from wind power in the state. Solar power is a frequent target of efforts to add surcharges to make it less affordable.

The industry is exercising its influence at a particularly delicate moment. Climate change is accelerating and global momentum has grown to deal with the crisis, culminating in last year's landmark Paris Agreement. But the U.S. elected a new president, Donald Trump, who has signaled solidarity with the industry and yanked the United States out of the

Paris accord. The president has branded climate change a hoax and installed officials at the Environmental Protection Agency and other key departments dedicated to expanding fossil fuel production and undoing Obama-era efforts to curb greenhouse gas emissions.

And the man at the center of the investigation by the attorneys general, Exxon's Tillerson, is now Secretary of State, a post giving him authority over climate diplomacy.

From Fanfare to Rounds of Legal Warfare

With great fanfare, New York's Schneiderman announced the formation of AGs United for Clean Power in March 2016. The coalition was formed after investigative articles first by InsideClimate News and later the Los Angeles Times disclosed that Exxon learned from its own scientists about the consequences of climate change as early as 1977, but chose to curtail its research and eventually worked to deny the findings.

The attorneys general, Schneiderman said at a news conference, were "dedicated to coming up with creative ways to enforce laws being flouted by the fossil fuel industry and their allies in their short-sighted efforts to put profits above the interests of the American people and the integrity of our financial markets."

Healey said, "Fossil fuel companies that deceived investors and consumers about the dangers of climate change should be, must be held accountable." Exxon would later charge that those comments indicated Healey had prejudged the outcome of her investigation.

At their side were five other attorneys general and staff representing 10 more, from California to Rhode Island.

Former Vice President Al Gore added star power to the assembly.

Despite the showy news conference, the attorneys general were unprepared for what was to come, and they made tactical errors at the start that gave Exxon ammunition for its counter-attack. David Vladeck, a Georgetown University law professor and former director of the Bureau of Consumer Protection at the Federal Trade Commission, said the group had too little understanding of each member's commitment. And they allowed the tiny Virgin Islands a prominent role that turned out to be a mistake, he said.

The morning of their news conference, the attorneys general met privately with climate activists, handing Justin Anderson, one of Exxon's lawyers, an opening to argue to Kinkeade that the coalition, all Democrats, was abusing the law enforcement system to seek a political goal. Kinkeade later seized on that argument, asking why the attorneys general wouldn't want to assure the public that their actions "lacked political motivation and were in fact about the pursuit of justice."

Eric Soufer, a spokesman for Schneiderman, said engaging a variety of sources is part of the investigative process. "Like all prosecutors' offices, we speak with a diverse array of stakeholders on a range of matters, all the time," he said. "It's part of our job."

Peter Frumhoff, director of science and policy for the Union of Concerned Scientists, said he briefed the attorneys general at the meeting on Exxon's "deception and the impact of climate change that might have been averted if they had taken action."

At the press conference, Attorney General Claude Walker of the Virgin Islands stood defiantly at the lectern and announced the fight was on with Exxon.

"It could be David and Goliath, the Virgin Islands against a huge corporation, but we will not stop until we get to the bottom of this," Walker said. He soon issued sweeping subpoenas to Exxon and the Competitive Enterprise Institute, a libertarian think tank aligned with and heavily financed by the industry.

Exxon and CEI almost immediately started filing a series of motions to have the subpoenas dismissed. They didn't play defense for long. Within weeks they filed lawsuits of their own against Walker, alleging he was violating their First Amendment right to disagree with the prevailing scientific consensus that climate change is overwhelmingly driven by burning fossil fuels.

Faced with a legal battle that might go on for years and the concern that losing an early round could undermine the larger investigations by New York and Massachusetts, Walker withdrew his subpoena of CEI and suspended his investigation of Exxon.

Sam Kazman, the general counsel for CEI, said Trump's election and his installation of Scott Pruitt, a climate change denier, as head of the EPA show that Schneiderman and the other members of what Kazman called "AGs United for More Power," badly misinterpreted the national attitude on the issue. "It's hard for Schneiderman to conclude that we were deceived when that skepticism is shared by the EPA under the new administration," he said.

Neither Walker nor lawyers representing the Virgin Islands would discuss the case or its dismissal. Donna Christensen, a

St. Croix resident who served 18 years as the islands' non-voting representative in the U.S. House of Representatives, said the territory relies on Congress for much of its budget.

"We are in a very fragile financial situation," she said. "We don't need to make any enemies."

In retrospect, said Michael Gerrard, who teaches climate law at Columbia Law School, the coalition had started a fight that was far greater than some were prepared to wage.

Exxon initially cooperated with New York, surrendering more than two million pages of records. But the company balked at a documents request from Healey and took its argument to Kinkeade, who allowed Exxon to question attorneys general and climate activists in its search for evidence of a conspiracy.

When Kinkeade, a Republican appointed by President George W. Bush, sent the case to New York, he laid out Exxon's allegations in detail, presumably a road map for his successor to follow. But Caproni, appointed by President Barack Obama, said at a hearing, "I have a different view of this case than Judge Kinkeade." When one of Exxon's lawyers sought a discovery order like the one Kinkeade granted to allow the company to try to prove that the attorneys general were improperly colluding, Caproni shut him down. "Gimme a break," she said.

Carroll Muffett, president and chief executive of the nonprofit Center for International Environmental Law, was one of those subpoenaed by Exxon.

"I knew from the very beginning that Exxon would eventually come after us hard and it would be a dirty fight,"

Muffett said. "They are fabricating conspiracy theories to deflect attention from the truth."

The Rockefeller Brothers Fund and the Rockefeller Family Fund, philanthropic organizations that promote environmental causes among their numerous programs, also were the targets of Exxon subpoenas. Both organizations are among donors that support ICN.

To defend itself, Exxon argued in the subpoena that it needed to know whether Muffett's and other non-governmental organizations influenced the attorneys general's decisions to open investigations.

That legal strategy prompted Sarah Labowitz, a New York University scholar who served nearly three years on Exxon's External Citizenship Advisory Panel, to resign.

Labowitz, who until recently was co-director of the Center for Business and Human Rights at the NYU Stern School of Business, said she had enough when the company went after Muffett and others unrelated to the two state investigations.

"I am disappointed that instead of examining its own record and seeking to restore a respected place for itself in the public debate, Exxon has chosen to turn up the temperature on civil society groups," Labowitz wrote in her resignation letter.

Watching the New York and Massachusetts investigations being sucked into a legal quagmire are the other attorneys general who were part of the coalition.

Maryland Attorney General Brian Frosh said, "I'm sure anybody who is considering moving ahead with an investigation or action would take into consideration the ferocity of what they've seen so far of Exxon's defense," he said. In addition to the federal case, the company has filed suits in state courts in New York and Massachusetts, seeking to quash the probes.

Exxon's Tactic of Delay

Exxon battles not only so it might profit while cases inch through the courts, but also to buy time until a more favorable political environment comes along.

For more than a decade, the company fought a lawsuit filed by the state of New Jersey to recover $8.9 billion, alleged damages from more than a century of pollution caused by Exxon refineries.

Brad Campbell was New Jersey's environmental commissioner when he authorized the lawsuit in 2004. He said in an interview that litigation became the last resort after Exxon refused to negotiate a settlement with the state as other companies, such as Chevron, had done.

Exxon lost a number of key battles that established its liability but nevertheless continued to fight, year after year, through the administrations of four governors.

"Exxon really stands in a class by itself in terms of a scorched earth approach to environmental claims of any sort," said Campbell, who is now president of the Conservation Law Foundation, a New England-based environmental organization that also has sued the company.

"They litigate more aggressively and often devote resources to fighting claims that aren't rational given the cost of resolving them in a constructive way."

But Exxon delayed long enough to get a favorable outcome. In 2015, at the direction of Gov. Chris Christie, who was campaigning for the Republican presidential nomination, New Jersey accepted a settlement of $250 million, or about 3 cents on the dollar.

Similarly, Exxon has been resisting for years a proposed federal pipeline safety fine of $2.7 million for a spill that dumped 210,000 gallons of heavy Canadian crude oil into the streets of Mayflower, Arkansas, in March of 2013. The fine for the spill, which sickened residents, forced evacuations and rendered some homes uninhabitable, would represent a tiny fraction of the company's 2013 revenues of more than $420 billion.

An Ally in the White House, but Much Explaining to Do

Exxon has enjoyed the consistent support of many Republican attorneys general and Republicans in Congress. That includes Rep. Lamar Smith, chairman of the House Science Committee and a climate change denier, whose largest campaign contributions last year came from the oil and gas industry.

The Texas Republican subpoenaed the attorneys general investigating Exxon, threatened subpoenas for the other coalition members and subpoenaed non-governmental environmental organizations that had supported the coalition.

Smith said during a news conference that, "The attorneys general have appointed themselves to decide what is valid and invalid regarding climate change. Attorneys general are pursuing a political agenda at the expense of scientists' rights to free speech."

Smith remains in a standoff with the environmental organizations and two attorneys general, who are refusing to comply with the subpoenas.

Unlike Smith, Exxon's legal team doesn't deny climate science. In fact, Exxon now acknowledges the risk of burning fossil fuels and has expressed support for a carbon tax. Echoing an argument he made to defend Philip Morris more than a decade ago, Wells says the company should be judged by its current behavior, not what it might have done in the past.

As New York investigators were digging recently through the mountain of documents Exxon turned over at the start of the investigation, they uncovered Tillerson's Wayne Tracker alias.

Now Exxon has a lot of explaining to do, said Wendy Jacobs, a professor of environmental law and director of the Harvard Law School Emmett Environmental Law & Policy Clinic.

The emails are a "scandal" that put "Exxon and its credibility and forthrightness in a bad light," said Jacobs, who served in U.S. Department of Justice's Environment Division.

"What will really derail Exxon is if there is a smoking gun buried in the Wayne Tracker emails. That remains to be seen," she said.

Vladeck, the Georgetown University law professor, said he thinks that the attorneys general lost time and resources with their initial missteps. But, he added, "The investigators have recovered, and the investigations are getting back on track."

IT TOOK THIS COAL MINER 14 YEARS TO SECURE BLACK LUNG BENEFITS. HOW COME?

Despite the federal Black Lung Benefits Act, 70 percent of benefit awards are challenged by an industry that hates to lose or pay.

By David Hasemyer
September 7, 2017

WISE, Virginia—Retired coal miner Bethel Brock proudly shows the vegetable garden where he grows potatoes, onions, corn, butter beans and other greens for the dinner table. He's in that patch nearly every day hoeing, shoveling—and struggling just to breathe.

Brock, a soft-spoken 77-year-old, has black lung disease from decades working in the coal fields of Virginia.

When his chest constricts and he becomes so winded that he can no longer stand, he slowly folds his body like a collapsing accordion, feeling more dread than discomfort.

"I have to stop right where I am and sit in the dirt," he said. "It feels like I'm suffocating."

Brock was diagnosed in 1982 with the early stages of black lung, a progressive illness common in coal miners. It eventually robs the lungs of the ability to hold air. In 2003, he began battling the Westmoreland Coal Company, its insurance carrier and their legion of lawyers and doctors for black lung benefits.

The company successfully fought him 10 times by maintaining he is not seriously stricken with the incapacitating disease despite nearly two dozen findings to

13

the contrary by his doctors—and one by a doctor hired by the company, which it did not disclose to a U.S. Labor Department examiner deciding his case.

Earlier this year, it looked as though Brock would finally win when the examiner ruled him eligible for benefits. But several months later, she inexplicably reopened the case and reversed herself on a technicality: The latest X-ray documenting the progress of his disease hadn't been stamped with the date it was taken. Once again Brock had lost in a system, created by federal law, designed to be adversarial and seemingly tilted in favor of industry.

But when InsideClimate News began asking questions, there was yet another surprising development: A supervisor decided the latest decision was "in error" and notified Brock he would be getting benefits after all. A U.S. Department of Labor spokeswoman then asked if ICN would be changing the focus of this story about the black lung benefits system.

Jennifer Grafton, Westmoreland's chief administrative and legal officer, said she was bound by privacy issues and could not talk about Brock's case. Colleen Smalley, the district director who overruled the examiner, refused to answer questions and hung up on a reporter.

Brock's 14-year fight with Westmoreland is typical of the David-and-Goliath battles coal miners face when seeking black lung benefits, which are paid by the company and include a monthly stipend and medical care for pulmonary illnesses. While thousands of miners and their survivors can struggle for decades to get benefits, coal companies and their insurers use every legal defense allowed under the Black Lung Benefits Act to deny them.

The companies engage high-profile law firms that in turn employ a cadre of doctors, setting the stage for protracted battles of conflicting medical opinions. Miners, meanwhile, are often overmatched because few lawyers will take their cases. Attorneys are barred from charging plaintiffs fees, meaning they receive modest compensation, and only if they win.

Over the last decade, 52,537 miners have applied to the Labor Department for black lung benefits. The department determined that only 7,252, or about 14 percent, were eligible, according to its data. The industry then challenged 70 percent of those claims, often denying the presence of the disease. It prevailed more than half the time.

In other words, of more than 52,000 claims, fewer than one in 10 was granted a disability award, despite the legal presumption that a miner with 15 years or more of service in the mines with lung problems has black lung as a result of his work. While rules were changed in 2000 to help level the playing field for miners, advocates, lawyers and statistics show companies still winning.

"The industry's tactic seems to be to just keep appealing until the miner or miner's wife dies or gives up," said Shannon Bell, an associate professor of sociology at Virginia Tech who has studied the impacts of the coal industry on the people of Appalachia.

This is just one front in wide-ranging battles by fossil fuel companies of all types to protect their bottom lines and their central role in providing America's energy.

"The fossil fuel industry thinks about the world differently than we do," said Brian O'Neill, who led a courtroom fight against Exxon on behalf of Alaskans devastated by the

Exxon Valdez oil spill. "It's about protecting the company and invoking a kind of immunity that the world depends on their product. They are nation states and they decide what is fair treatment."

The conflicts rage in Washington, where the industry works to undermine environmental regulation, climate policy and science; in state capitals, where it seeks to overturn laws supporting renewable energy; and in the courts, where, for example, ExxonMobil tries to foil fraud investigations by state attorneys general. Many of these skirmishes, like thunder and lightning in distant storm clouds, seem far removed from the lives of everyday folks.

But sometimes, such as when an energy company sues a Texas couple and their environmental consultant for defamation for claiming its fracking operations contaminated their residential well, or when a Louisiana family faces confiscation of their land after they sue over leaking oil pipelines, the fight can get very personal.

Nothing seems quite as personal as the struggles of Brock and fellow black lung victims to get benefits, including health care that can mean the difference between life and death.

Brock's Story: A Long March Toward Disability

Bethel Brock is slender with a soft, deliberate voice. He dropped out of Kelly High School at 15 to go into the mines with his dad to help support his family. In 1958, his father was killed in a traffic accident, and Brock, just 18, was trapped in a life underground, working for $8 a day to help his widowed mother.

"Coal jobs is all there were in these mountains," he said. "There was nothing else to do, especially with no education."

Brock didn't like mining, so he taught himself to be a mechanic and began working on mine equipment and driving coal trucks. He was out of the coal shafts, but it wasn't steady work, and he wasn't earning much.

When he married at age 20 and became a father, the mines beckoned again.

In 1968, at 28 with three children and another on the way, he hooked on with Westmoreland at its Wentz mine, hard by the Virginia-Kentucky border. The company had operations scattered across the country, and Brock did his small part in mining billions of tons of coal worth billions of dollars.

He called the Westmoreland job a Godsend because of the pay and benefits. One of the first things the company did was cover the hospital bills for his fourth child.

In the deep mines, it was impossible not to breathe coal and rock dust so dense it becomes an opaque curtain, like a widow's veil. At the end of the day, Brock emerged so black that, when he smiled, his teeth shined brilliant against his face.

"I knew what I was breathing, but I didn't have a choice," he said. "When you have a family to feed, you think about that first and not about the danger."

Black lung—known medically as coal workers' pneumoconiosis—can lurk for years and sometimes decades in a miner's body before symptoms emerge.

Coal dust gradually builds up in the lungs and cannot be removed by the body. Over time, it leads to inflammation, scarring and loss of tissue and elasticity so that the lungs cannot fill with air. They turn from a rosy pink to black as the disease progresses.

Brock was feeling fine the day in 1982 he had a routine physical and chest X-ray as part of a National Institute for Occupational Safety and Health program. There was nothing that suggested the results would be anything but negative as they had been for years.

But this time, the X-ray showed spots on his lungs. He remembers looking at the X-ray and thinking the white spots looked like BB holes against a black target. It was the first sign of the irreversible disease that would define the rest of his life.

The diagnosis was stunning—and numbing: Would the disease kill him? How could he care for his family? What if he'd never become a miner? What now?

"When you hear something like that, it changes your life," he said.

Brock was in his early 40s and the father of seven children, four girls and three boys between the ages of 5 and 18.

"At a moment like that, you think about a lot of things," he said. "But more than anything else, I thought about how long could I keep providing a paycheck for my family."

The diagnosis would ultimately be his ticket out of the mine.

Federal laws designed to safeguard miners mandate that someone diagnosed with black lung be given opportunities to work above ground.

Backed by the United Mine Workers of America, a union Brock lauds for its tireless fight on behalf of miners, he was given a job in 1985 on the surface processing mined coal.

For the next 14 years, he still breathed coal dust, but less than in the mines. And each year, X-rays showed a slow progression of the disease; an agonizing march toward total disability—and maybe death.

Between 2001 and 2015, 8,348 deaths were attributed to black lung, according to data from The National Institute for Occupational Safety and Health. More than 70 percent of the deaths were among people older than 75.

Working for Westmoreland was a good job, though Brock says he now believes his loyalty and hard work meant nothing to the company. That was clear to him when it began fighting his black lung claim.

"They turned against me like I was nothing but a used piece of equipment to be put on the junk pile," he said.

While Grafton, the Westmoreland official, wouldn't comment on Brock's case, she defended the company's approach to black lung claims.

"Each claim is handled on a case-by-case basis," she said. "There certainly isn't a blanket policy to deny every claim and make everyone fight."

She said Westmoreland relies on outside experts she declined to identify, including doctors, to advise about each case. "Once we have that information, we make a decision on how to proceed," Grafton said.

A 'Cottage Industry' Against Coal Miners

Coal companies and their insurance companies have superior resources to mount sustained legal resistance at a time when recent federal research shows black lung is affecting a higher percentage of miners than at any time since the 1970s.

Miners, advocates and attorneys immersed in prolonged black lung legal battles say the system was almost perversely set up to deny benefits to black lung victims.

"There is a cottage industry of lawyers and doctors working against the miners, to do nothing but reject black lung claims," said Dr. Joe Smiddy, a Kingsport, Tennessee, pulmonologist who has been treating coal miners for decades.

"What we are seeing is the industry denying the science and reality about the dangers of black lung as a disease in the same way the tobacco industry denied the danger of smoking," he said.

The companies often spend tens of thousands of dollars on lawyers and expert medical witnesses to attack a single claim. Although the monthly benefit pays $651 for a single miner or $976 for a miner and spouse, it's the requirement that all of a miner's medical expenses associated with black lung be covered that spurs such ferocious opposition.

The medical benefits include everything from prescription inhalers to lung transplants, which can cost $1 million or more.

Since 2000, incremental changes in federal regulations have somewhat leveled the field for applicants. Three of the most significant changes include limiting the number of X-ray exams that can be presented as evidence to two, accepting the presumption that 15 years of work underground is the cause of any respiratory illness, and requiring that medical evidence developed by coal companies be turned over to the miner.

The limit on X-rays means companies can no longer use their resources to overwhelm miners with expert opinions.

The 15-year presumption shifts the burden of proof from the miner to the company.

The disclosure requirement ends a practice under which lawyers for coal companies withheld medical evidence unfavorable to their clients.

Stephen Sanders, a lawyer and director of the Appalachian Citizens Law Center in Whitesburg, Kentucky, who has represented miners in black lung cases for 25 years, welcomed the rule changes but said he has not seen much change in the success rate for his clients.

"These operators avail themselves of every legal avenue available to them," he said. "Opposing claims is an automatic response that doesn't depend on medical evidence or the obvious suffering of a claimant."

After Brock's black lung disease had progressed to the point where he said he became disabled, he filed a claim with the U.S. Department of Labor's Division of Coal Mine Workers' Compensation in 2000. He was denied and didn't re-file until three years later when his symptoms worsened.

The second claim was approved in 2003, triggering the 14-year battle with Westmoreland.

Westmoreland's argument rested on alternate interpretations of Brock's X-rays by doctors hired by the company. They said there was no indication he was suffering from black lung disease.

"Mr. Brock does not have any respiratory disability from any cause including coal workers' pneumoconiosis," read notes written in 2006 by a doctor hired by the company.

The doctor went further, suggesting Brock's health problems were related to heart disease and high blood pressure.

"Both of these conditions of the general public at large are unrelated to the inhalation of coal mine dust and coal workers' pneumoconiosis," the doctor wrote.

Another doctor hired by the company said Brock was healthy enough to go back to work in the mines.

Companies routinely offer different explanations for miners' symptoms, blaming them on smoking, tuberculosis, pneumonia and even bat guano inhaled from the mines.

Some have suggested the symptoms are merely psychosomatic.

But no less than seven doctors since 2003 have looked at the spots on Brock's X-rays and diagnosed black lung disease, several of them on multiple X-rays.

"Mr. Brock has evidence of an occupational pneumoconiosis which occurred as a direct consequence of his prior coal mining employment," according to one of his doctors in 2004.

Brock's most recent X-ray, in July by Dr. Smiddy, showed the disease continuing its deadly progression.

It's difficult to predict Brock's future, but Smiddy said miners exhibiting his symptoms eventually require full time oxygen and often wind up in wheelchairs.

Yet, time and time again, Westmoreland prevailed after labor board claims examiners agreed with assessments by company doctors that Brock did not have black lung.

Such decisions sometimes seem influenced by the resumes of the doctors the companies can afford.

"It puts the miners at a disadvantage fighting against a company that has substantially more resources," said Jeremy O'Quinn, Brock's attorney.

Wrongful Denial and Charges of Fraud

Among the first doctors Westmoreland hired to review and diagnose Brock's condition was a prominent, Harvard-trained radiologist at the Johns Hopkins University School of Medicine black lung unit.

Paul Wheeler and two of his colleagues, all board certified in the specialty of black lung radiology, reviewed Brock's X-rays and determined he was not suffering the effects of the disease.

The diagnosis fit a troubling pattern.

Wheeler had not confirmed a single case of severe black lung in more than 1,500 cases he evaluated between 2000 and 2013, according to an investigation by reporter Chris Hamby of the Center for Public Integrity. Hamby's series, "Breathless and Burdened," won a Pulitzer Prize in 2014.

The disclosure led the U.S. Department of Labor to notify about 1,100 coal miners that their claims for black lung benefits may have been wrongly denied because of Wheeler's findings. It also resulted in Johns Hopkins closing its black lung unit and led to a class action lawsuit against the university and Wheeler.

The 2016 complaint accuses Johns Hopkins and Wheeler of defrauding coal miners out of black lung benefits using faulty diagnoses.

According to the lawsuit, Wheeler intentionally deviated from an international standard for establishing black lung and substituted his own non-conforming standards for reading X-rays. In that way, "Johns Hopkins and Dr. Wheeler were successful in denying lawfully earned federal benefits to injured coal miners," the compliant alleged.

Johns Hopkins represented the "gold standard" in health care, and, consequently, its credibility carried great weight, said Jonathan Nace, one of the Washington attorneys representing the miners.

Kim Hoppe, director of public relations and corporate communications for Johns Hopkins, said she could not comment because of the litigation, but she noted the institution has not resumed its black lung program, which was suspended in the wake of the disclosures regarding Wheeler and his colleagues.

Wheeler could not be reached for comment.

In response to the lawsuit, Wheeler and Johns Hopkins contended they are immune from responsibility because expert witnesses in adversarial proceedings are protected by a litigation privilege.

"Disappointed claimants cannot sue one of the employer's expert witnesses in court to recover additional benefits not awarded in the administrative proceedings," according to the rebuttal.

Brock and his attorneys also found themselves up against one of the largest black lung law firms in the country. The Jackson Kelly partnership of Charleston, West Virginia, has been a go-to law firm for decades for coal companies.

But Jackson Kelly's zealous advocacy has drawn charges of fraud and led to the suspension of one its lawyers by the West Virginia Supreme Court.

In 2011, Douglas Smoot lost his law license for a year because he failed to provide evidence to a miner that would have helped the man prove that he had black lung. The West Virginia State Bar Office of Disciplinary Counsel found Smoot "engaged in misconduct by improperly withholding" critical evidence favorable to the miner while representing Westmoreland.

Smoot represented Westmoreland against Brock.

The law firm also has faced allegations in a class action lawsuit that it withheld information for decades that its own doctors had confirmed the most severe form of black lung in miners.

"In developing evidence in the black lung cases, Jackson Kelly attorneys have knowingly submitted evidence that misrepresents the facts, including, in some cases, the opinions of their experts," according to the 2009 lawsuit filed in circuit court in Raleigh County, West Virginia.

There is no allegation of wrongdoing by Jackson Kelly or Smoot in Brock's case. The 2009 lawsuit was recently dismissed with details kept confidential. Jackson Kelly and Smoot did not respond to a call for comment.

In March, Brock got the news he'd been hoping for after 10 appeals. The Labor Department determined that his condition had worsened enough that he qualified for black lung benefits.

Brock had filed his latest appeal with an X-ray and doctor's opinion that he was suffering from complicated black lung disease, the same diagnosis he had been receiving for years.

Strangely, Westmoreland's new law firm, Bowles Rice, which had taken over the case in 2013, didn't file a response. After a deadline for submitting evidence passed, the reason for the silence became clear. The firm disclosed to Brock and O'Quinn the results of an X-ray exam its own doctor had performed: Brock had black lung.

"There it was," Brock said. "The company's doctor was saying it's black lung."

Even without that evidence, which would have clinched the case, Labor Department Examiner Debbie Weyandt ruled in Brock's favor. But three months later, she changed her mind. Brock's latest X-ray didn't have a date stamped on it.

O'Quinn said he tried to submit Westmoreland's black lung diagnosis after the claims examiner rejected Brock's X-ray, but she refused to consider it.

When InsideClimate News asked the Labor Department to explain what prompted Weyandt to reopen the case and reverse herself, a reporter was told that privacy laws prevent discussion of individual cases. Brock waived his right to privacy and authorized the government to release his records and discuss his case with ICN.

Smalley, the district director, then issued a new order. "I find the X ray evidence supports the presence of complicated pneumoconiosis," she wrote, deciding it was "irrebuttable" that Brock was entitled to benefits.

Smalley told Brock she acted on her own initiative and wouldn't discuss his case with ICN.

Brock said he hopes this is the end of his long fight with Westmoreland, which now has another opportunity to appeal.

"You never know what this company is going to do," he said. "They have fought me so hard, nothing would surprise me."

A Miner Led to a New Calling

Westmoreland closed its Virginia operations in 1995 and Brock was laid off. He used the severance and free time to return to school at age 55. He earned his General Education Diploma and certification in heating and air-conditioning repair from Mountain Empire Community College. When black lung prevented the strenuous work of a repairman, he went on to the University of Virginia, where he earned a bachelor's degree in political science and credentialing as a paralegal.

For the next eight years, he worked for attorney Joe Wolfe, helping miners navigate the complex and intimidating maze of regulations surrounding black lung claims.

"It gave me the chance to help miners who were lost and confused by the system," he said.

He no longer works full time for the law firm, though he occasionally assists miners with claims from a home office where a poster of John L. Lewis—the firebrand former president of the United Mine Workers of America—hangs on one wall.

Although Brock lives comfortably on Social Security, his United Mine Workers pension and income from rental property, there are many miners who barely scrape by and would have a better quality of life with black lung benefits.

Along with a handful of other former miners, Brock has formed the Black Lung Association of Southwest Virginia, a small grassroots organization he hopes will grow to become a thundering voice heard in state capitols and Washington.

At a recent meeting in Cleveland, Virginia, to recruit members, Brock and Dean Vance, the organization's secretary, listened to the stories of dispirited miners who'd surrendered to the overwhelming odds they'd faced in their cases.

"We're never going to win," said Burl Rhea, who'd spent 29 years as a miner.

"Don't give up. You have to fight," Vance told the group. "When you say, 'I quit. I ain't gonna fight anymore,' it just makes the coal companies happy."

Brock said he wants this story to be about more than his experience. He wants it to be about every miner who has ever faced the overwhelming odds of fighting for black lung benefits.

"I am just one example of how the big coal companies use their money and influence to stand in the way of the benefits due miners who give their lives for coal company profit," he said.

HOW FOSSIL FUEL ALLIES ARE TEARING APART OHIO'S EMBRACE OF CLEAN ENERGY

With scare studies, policy drafts and political donations, industry groups turned Ohio lawmakers against policies they once overwhelmingly supported.

By Brad Wieners and David Hasemyer
October 29, 2017

COLUMBUS, Ohio—On March 30, Bill Seitz, a charismatic Republican, took to the floor of the Ohio House to make a case for gutting a 2008 law designed to speed the adoption of solar and wind as significant sources of electricity in the state. The law, he warned, "is like something out of the 5-Year Plan playbook of Joseph Stalin." Adopting a corny Russian accent, he said, "Vee vill have 25,000 trucks on the Volga by 1944!'"

Nine years before, Seitz and his colleagues, Republicans and Democrats alike, had voted overwhelmingly for the measure he now compared to the work of a Communist dictator. It made Ohio the 25th state to embrace requirements and inducements to lure utilities away from coal, a major contributor of the gases fueling global climate change. Studies suggested the law would help create green energy jobs and boost the Ohio economy—and it has.

Now, Seitz said, it was obsolete. Natural gas, rapidly displacing coal, was the resource Ohio ought to foster, he said. He also argued the law gives an unfair advantage to wind and solar when the state's last nuclear plant is fighting for its life. Most important, Seitz insisted, the government had no business telling anyone what kind of energy to buy. By the time he was done, he had secured a veto-proof majority to undo key parts of the law.

What happened to turn lawmakers so decisively against a statute they'd adopted 93-to-1 less than a decade ago?

The answers begin with the 2008 financial crisis, which hit Ohio hard and greatly depressed energy demand, and they include the shale gas boom, which benefited Ohio producers and made coal uncompetitive.

But there's more to the story, too.

As fossil fuel interests mobilized at the national level to fight proposals to mitigate climate change that would undercut their profits, they made Ohio a priority for fighting clean energy policy at the state level. Beginning in earnest in 2011, a network of coal companies, utilities, think tanks, nonprofit foundations and political action committees coalesced to roll back Ohio's alternative energy initiatives.

Industry-supported think tanks provided highly questionable research purporting to show big job losses. An industry group claiming to represent consumers—and accused of using fraudulent tactics before regulatory agencies—advised Seitz's staff on how to water down the definition of alternative energy. And industry sources donated to the campaigns of state politicians, like Seitz, who've kept the repeal-and-replace bills coming, even after Republican Gov. John Kasich vetoed a similar effort.

This network includes Americans for Prosperity, a foundation funded by the energy magnates Charles and David H. Koch; the Heritage Foundation, a Washington-based advocacy group known for its criticism of climate change science; and the American Legislative Exchange Council (ALEC), another conservative nonprofit in

Washington with Koch ties that frequently spoon-feeds draft legislation to state politicians.

Seitz is on ALEC's national board of directors, but he bristles at the suggestion that he relies on the council for guidance. "ALEC doesn't drive me," he told InsideClimate News. "If anything I drive ALEC." Either way, in 2012, ALEC adopted an "Electricity Freedom Act" that reads like a declaration of war against the kind of energy rules on the books in Ohio and calls for the effort to reject them that Seitz leads.

The Rise of Clean Energy Scare Studies

Once a major coal producer, Ohio has a lesser-known history with fuel-free electricity. In the winter of 1887-88, Charles F. Brush constructed the world's first electricity-generating wind turbine on a 60-foot iron mast behind his Cleveland mansion. Brush held more than 50 patents, and his company joined with several others to form General Electric. In 1984, physicist Harold McMaster first demonstrated in his Toledo basement the potential for thin, non-silicon-based films to capture the sun's energy. Glasstech, now called First Solar, has led the conversion of Toledo into a solar manufacturing hub. The University of Toledo trains engineers for the many photovoltaic companies in the region, and, in 2007, it added the Wright Center for Photovoltaics Innovation and Commercialization to transfer what's in its labs to nearby factories.

"Ohio has a legitimate claim to being a leader in renewable energy, which is what makes the effort to stymie renewables here all the more ironic," said Bill Spratley, former executive director of Green Energy Ohio, an advocacy organization in Columbus. He said if the Seitz bill became law, "we might end up supplying everyone else with these wonderful technologies while we don't get to benefit from them ourselves."

Spratley testified against Seitz's bill. He noted that there are roughly 80 companies making parts for wind power systems in Ohio and scores of engineers and sales representatives installing solar, "and those jobs aren't going anywhere," he said. "Those are local jobs."

Ten years ago, more Ohio politicians embraced Spratley's message. In 2007-08, as the "Energy, Jobs and Progress" plan made its way toward law, energy demand was strong, prices were expected to remain high, and awareness of coal's contribution to climate change had peaked. The law committed Ohio to cutting energy consumption by 22 percent by 2025 and diversifying sources so that 12.5 percent of its electricity would come from alternative energy sources—geothermal, biomass, wind, solar.

Demand and prices fell with the recession and the shale gas boom, but the promise of more jobs and less global warming continued to resonate. Or it did until studies started showing up that warned that the law would do more harm than good.

In 2011, the former American Tradition Institute, now the Energy and Environment Legal Institute, commissioned an analysis of the economic impacts of Ohio's energy standards. Based in Washington, the EELI receives funding from Peabody Energy and Arch Coal, the two largest suppliers of coal in the United States. The Beacon Hill Institute, then an affiliate of Boston's Suffolk University and a frequent publisher of papers critical of climate-related policies, undertook the study.

In its report, Beacon Hill forecast that the Ohio energy standards would result in the loss of 9,750 jobs, wipe out $1.1 billion in disposable income every year, and force Ohioans to pay $8.6 billion more for electricity in 2025. The authors, among them BHI's executive director, David G. Tuerck, reached this conclusion based on an economic model that ignored the rapidly declining cost of electricity generated by wind and solar. Beacon Hill later lost its Suffolk University affiliation because, a university spokesman told The Boston Globe, its research lacked rigor and tended to reach conclusions sought by its underwriters.

On April 13, 2011, ALEC sent the Beacon report to all its members in the Ohio legislature, warning that the energy rules were "a blueprint for economic misery."

Citing the study, state Sen. Kris Jordan introduced a bill five months later to repeal the alternative energy standards. The measure didn't make it out of committee, but Jordan, Seitz and others kept at it, until, in 2014, they managed to secure a two-year freeze on meeting the annual benchmarks established under the 2008 law.

More scare studies followed. In 2015, the Institute of Political Economy at Utah State University published a paper suggesting Ohio's energy rules would kill off 29,000 jobs. Last spring, the Buckeye Institute, a free-market think tank across the street from the Ohio Statehouse in Columbus, modeled four economic scenarios; the rosiest estimated that the renewable and energy efficiency standards would cost Ohio 6,800 jobs and $806 million in GDP by 2026.

"Those studies are completely bogus," said Terrence O'Donnell, a lawyer representing renewable energy developers at Dickinson Wright, a Columbus law practice. "The Utah one is the most ridiculous, because it blames everything that happened to the economy during the recession on the renewable portfolio standards. It's laughable. There's not one policy maker I know of who still refers to it." The Buckeye Institute, he added, assesses "a fictitious Ohio law, not the one on the books." For example, the Buckeye economists assume "that the cost containment provisions in the law simply won't kick in if prices for renewable energy rise. But they don't say why." And the price of solar electricity hasn't just fallen a little bit; it's less than half of what it was three years ago.

Several other studies have concluded that Ohio's energy rules have, on the contrary, created jobs and improved the economy. One, published by Ohio State University's Center for Resilience, found that in 2012 alone, the law stimulated a modest .04 percent, or $160 million, in GDP growth statewide. According to the Ohio Environmental Council (OEC), the state added more than 2,300 renewable energy projects and 25,000 clean energy jobs since 2009. Not minus 29,000 jobs, but plus 25,000 jobs, and not based on a model, but on payroll and labor statistics.

Trish Demeter, the OEC's managing director of energy, said the 2008 law benefits the state in another way. Its energy efficiency provisions are saving money. For every dollar invested in efficiency, she said, "Ohio ratepayers are saving between $1.10 and $4.20 on their utility bills, depending on what kind of consumer they are." She said this data came from actual receipts, not models, and had been reviewed by the Ohio Consumers Council and the American Council for an Energy-Efficient Economy (ACEEE).

So whose numbers are right? On behalf of Gov. Kasich's office, Matt Cox, a Ph.D. from MIT and founder of Greenlink, a consulting practice in Atlanta, modeled three scenarios to try to determine if Ohio's energy rules were too aggressive. Overall, Cox found that the rules generated more jobs, more GDP and, in time, lowered electrical bills.

Seitz told InsideClimate News that Cox ought not to have factored public health impacts of air pollution into his study. Cox responded, "How can you do a cost-benefit analysis and not factor in a cost that is borne, not by the utility, but by everyone else?"

Dave Anderson is policy and communications manager for the Energy & Policy Institute, a watchdog group headquartered in San Francisco. He and his team note that all three of the organizations behind the Ohio scare studies—Beacon Hill, Utah State and Buckeye—have received donations from the Koch brothers or their affiliates. For example, the Buckeye Institute receives funding from the Koch-backed Donors Capital Fundand Claude R. Lambe Charitable Foundations, as well as directly from the Charles G. Koch Charitable Foundation.

Passionate Opposition to Wind Power

The bill that froze Ohio's energy rules in 2014 also created an Energy Mandates Study Committee to determine whose numbers were best and what to do next.

To staff the committee, Seitz turned to Sam Randazzo, general counsel for a consortium of electricity purchasers large and small called the Industrial Energy Users of Ohio. Randazzo has been deeply involved in shaping Ohio's energy policy for five decades and, like Seitz, opposes the 2008 standards.

On Aug. 8, 2016, as the freeze neared expiration and the EMSC was due with its findings, Seitz wrote to 10 utility, gas and coal lobbyists, as well as Randazzo, that "we should be meeting as a small group to figure out what that report is going to say." (The watchdog group EPI obtained this email through a public records request.)

The EMSC counted 12 elected officials, among them Seitz, as members. These 12 collectively received $830,000 in campaign contributions from utilities, oil and gas interests, and coal mining companies, according to an investigation by the National Institute on Money in State Politics. Contributions from electric utilities to Seitz more than tripled after he began trying to dismantle the state's renewable energy standards.

Randazzo disputed any notion that the committee was stacked by the fossil fuel industry. "This was a legislatively appointed body with representation by both Republicans and Democrats," he said. And, it was "absolutely open to all stakeholder positions."

In an interview, Seitz was blunt about whose interests he was looking out for. "Coal is moribund and crowded out by natural gas. If there's anyone I should have an affinity for it's the natural gas guys. ... The old saying is, you dance with who brung you to the dance. In Ohio, natural gas is who brung us to the dance."

Seitz despises wind turbines, and his dedication to rolling back Ohio's energy standards stems in part from his passionate opposition to wind power. In fact, the only reason he voted for the standards back in 2008, he said, was a promise by the Senate president that he could write the language on turbines. Over time, Seitz has been behind restrictions that have wiped out any large new projects. "There will be no wind farms!" he said, with satisfaction.

Turbines, Seitz said, take up too much room, don't work when the wind doesn't blow, and are not a good fit for his district, in the far corner of the state where it borders Indiana and Kentucky. His neighbors love spending time in their yards and don't want any turbines wrecking their "view sheds," or chopping up bats and birds, he said. And it doesn't help matters that "it seems to be a cabal of urban millennials who love wind [power] and want to inflict the damage on rural landscapes—stick it out there in the country where all the bumpkins can't do anything about it. That's not very nice. So I'm not a big fan of wind."

The final Energy Mandates Study Committee report was a gift to incumbent utilities and gas and coal interests. It recommended an indefinite extension of the freeze on renewable standards and more or less mirrored a bill Kasich vetoed on Dec. 22, 2016, saying the measure could undermine the state's improved business climate and prevent businesses and homeowners from saving money by saving energy.

Undeterred by Kasich's veto, HB 114, the bill that passed on March 30, contains much of the EMSC wishlist.

Seitz's staff also received advice from the Consumer Energy Alliance on expanding the definition of what counts as an alternative form of energy. The CEA is comprised of few consumers, operates out of HBW Resources, a Washington-based lobbying firm, and is backed by major oil and gas companies, such as Chevron, ExxonMobil, Marathon, Shell and Norway's Statoil. It's a lobbying group that has been challenged before the Federal Energy Regulatory Commission for submitting public comment letters from individuals who later said in sworn statements that they never signed them. The CEA withdrew a petition in Wisconsin after a local reporter found that some people who signed it actually opposed the utility rate changes CEA was pushing.

Seitz laughed at the notion that he is the face of a well-funded industry campaign. "As I've made clear 20 times, I find mandates philosophically obnoxious," he said. "Look, I'm an antitrust lawyer by background and training. Antitrust lawyers tend to favor competition and not be a big fan of monopolies, and tend to believe that government interference ought to be kept to a minimum."

His bill is now in the state Senate, where it may face an uphill fight. John Fortney, spokesman for the Republican majority caucus and Senate President Larry Obhof, said any legislation will have to be a compromise acceptable to both houses and Kasich. Kasich's office did not respond to repeated requests for an interview.

Andrew Kear, an assistant professor of political science and environment and sustainability at Bowling Green State University, said HB 114 can't survive without substantial changes. He said any measure will have to recognize that renewable energy is an economic driver in parts of the state. "They have to get beyond the false dichotomy that it's environment versus economy," he said.

INDUSTRIAL STRENGTH: HOW THE U.S. GOVERNMENT HID FRACKING'S RISK TO DRINKING WATER

A pivotal EPA study provided the rationale for exemptions that helped unleash the fracking boom. The science was suppressed to protect industry interests.

By Neela Banjeree
November 16, 2017

Most mornings, when his 7-year-old son Ryan gets up for school at 6:55, Bryan Latkanich is still awake from the night before, looking online for another home in some part of Pennsylvania with good schools and good water.

Six years ago, Latkanich signed on to let an energy company tap natural gas beneath his property by pumping water, sand and chemicals into rock formations, a process called hydraulic fracturing, or fracking. Soon after, Latkanich's well water got a metallic taste, he developed stomach problems, and his son one day emerged from a bath covered in bleeding sores. More recently, Ryan became incontinent.

Testing by state regulators and a researcher at nearby Duquesne University showed the well water had deteriorated since gas extraction started but no proof of the cause. The state recently began another round of testing.

Latkanich is a single parent. He's jobless and blind in his right eye from brain surgery. "I worry about my son getting sick, about my getting sick and what would happen to him if I did," he said. "I'm doing this all alone. And I keep asking myself, 'How do we get out?'"

For Latkanich and all those who believe their water has been tainted by fracking, there are few remedies. Congress took away the most powerful one in 2005, prohibiting the Environmental Protection Agency from safeguarding drinking water that might be harmed by fracking and even denying the regulator the authority to find out what chemicals companies use. That provision of the Energy Policy Act was justified by an EPA study about fracking into coalbed methane reservoirs, completed under the George W. Bush administration, that concluded that fracking posed no risk to drinking water.

Concerns about the study emerged from the outset, including a 2004 whistleblower complaint that called it "scientifically unsound." Now, InsideClimate News has learned that the scientists who wrote the report disagreed with the conclusion imposed by the Bush EPA, saying there was not enough evidence to support it. The authors, who worked for a government contractor, went so far as to have their company's name and their own removed from the final document.

At EPA, "there was a preconceived conclusion that there's no risk associated with hydraulic fracturing into coalbed methane. That finding made its way into the Energy Policy Act, but with broader implications," said Chi Ho Sham, the group manager of a team of scientists and engineers for The Cadmus Group, the Massachusetts firm hired to do the report. "What we would have said in the conclusion is that there is some form of risk from hydraulic fracturing to groundwater. How you quantify it would require further analyses, but, in general, there is some risk."

The fracking provision, widely known as the Halliburton loophole, after the oilfield services company once run by Bush's vice president, Dick Cheney, is among a host of exemptions to federal pollution rules that Congress and successive administrations have given oil and gas companies over the last 40 years.

Winning these exemptions is at the heart of a successful strategy by the fossil fuel industry and its allies in Washington to limit environmental oversight of companies' operations. As a result, oil and gas drilling and production are exempt from laws regulating hazardous waste, chemical-laced runoff from well sites and toxic air pollution from well equipment. Some exemptions, such as the Halliburton loophole, were justified by EPA studies whose findings were ignored or bent to political ends, according to documents and interviews with scientists, lawmakers and former regulators who have worked on federal rulemaking since the late 1970s.

The Cadmus study was not the first EPA report to have its science thwarted, and under President Donald Trump, it likely won't be the last. Current EPA Administrator Scott Pruittis a staunch ally of fossil fuels, and his agency is moving on several fronts to quash science that documents the oil industry's contributions to climate change and other forms of pollution, the first step to rolling back regulations, critics said.

"I've been caught off-guard by how fast and diverse the attacks are on scientists within the government and how science is used," said Gretchen Goldman, research director for the Center for Science and Democracy at the Union of Concerned Scientists.

The EPA did not respond to multiple requests for comment made over two months. Former EPA officials from the Bush administration involved with the study would not comment on the record. Cadmus also would not comment and referred inquiries to the EPA.

The consequences of loopholes built on disputed science have rippled through the country during the latest energy boom. Domestic production of oil and gas has surged, creating thousands of jobs and boosting company profits— and leading to thousands of complaints in states such as Pennsylvania, Texas and North Dakota that drinking water is being contaminated. But, in the absence of federal protections, there is only a patchwork of often-lax state regulations. If it were not for the Halliburton loophole, the EPA could have developed standards for the entire country. State rules could have been tougher, but not weaker, than the national standards, and if states failed to regulate effectively, citizens could have petitioned the federal government to intervene.

"My dream was to build houses on this land for my sons and their families when they grew up, and to start a truck farm when I retired," Latkanich said. "Now I'm just fighting a battle by myself against a billion dollar company."

Chevron Comes Calling: 'This Was a Godsend'

Before four Chevron Appalachia employees came calling in 2011 with promises of riches, Latkanich's life had crumbled. In 1998, he had used an inheritance to buy land from a farmer in Deemston, 35 miles south of Pittsburgh. Latkanich was a counselor at the Washington County jail, often working with murderers. His wife was a nurse at a state penitentiary. They bought the rural tract as a haven from their tough jobs and built a dream house on a hill, with a wide front porch overlooking a two-acre pond.

But by 2010, the marriage had ended. His wife had left for a nearby town with his two older sons. Latkanich underwent an operation to remove a benign brain tumor, which, because of its size and location, threatened his life. While he was in a coma, his girlfriend gave birth to Ryan. She was addicted to cocaine and opioids, and the newborn spent three weeks going through withdrawal. The state placed Ryan in foster care.

When Chevron Appalachia showed up, Latkanich was on disability. He had spent the year in a hospital bed in his dining room with failing kidneys, back problems and $150,000 in bills from lawyers handling his divorce and efforts to regain custody of Ryan. Chevron offered him $400 for each of his 33 acres and estimated thousands more in royalties a month once the gas started to flow. He signed on. "In my situation, when it looked like I could lose everything, this was a godsend," he said.

To Regulate or Not: Industry Gets Boost From Cheney

The two wells on Latkanich's property are among 1,655 that have been hydraulically fractured in Washington County since 2004. Halliburton fracked the first commercial well in the United States in 1949. Technology has improved over time, getting a big boost from more than $135 million in federal grants beginning in the 1970s to spur development of oil and gas in shale formations. In the 1990s, fracking was used to extract coalbed methane, or natural gas, touted then as the next great investment for the industry.

The EPA and industry long maintained that fracking did not need federal oversight under the Safe Drinking Water Act (SDWA). The EPA used the law to protect groundwater from other industrial activities, such as disposal of oilfield wastewater as part its Underground Injection Control (UIC) program. But the agency contended that fracking did not fall under the UIC program and state oversight was adequate.

That assertion was successfully challenged in 1997 when the Legal Environmental Assistance Foundation (LEAF) won a case against the EPA on behalf of an Alabama couple who said their well water had been contaminated by nearby fracking for coalbed methane. The LEAF suit alleged that federal oversight of fracking under the SDWA was needed because the process was in fact a form of underground injection and state regulation was insufficient.

LEAF's success scared the industry and politicians allied with it, said Hannah Wiseman, a law professor at Florida State University. They didn't want federal rules that would have required a UIC permit for each frack job, potentially slowing energy extraction and choking revenues.

In 1999, Sens. James Inhofe (R-Okla.) and Jeff Sessions (R-Ala.), longtime allies of the oil industry, introduced a bill to exempt fracking from the Safe Drinking Water Act. A year later, the EPA announced a study to determine if fracking into coalbed methane reservoirs affected drinking water.

Industry got a huge boost when Cheney, the CEO of Halliburton, became vice president in 2001. At the time, fracking was unknown to the broader public. But an energy policy task force Cheney helmed in spring 2001 highlighted fracking's potential, and it recommended a comprehensive exemption to the SDWA for all types of fracking, not just for coalbed methane. The EPA cautioned against an overly broad approach.

EPA Administrator Christine Todd Whitman wrote to Cheney on May 4, 2001, "I strongly suggest limiting the recommendation to the problem we know about—hydraulic fracturing for coalbed methane. Otherwise, before the (coalbed methane) study is completed, we are potentially walking into a trap because we don't yet know the environmental consequences of the broader exemption, or why it is needed."

A draft version of the coalbed methane report was released in 2002 for public comment. Industry and environmental activists alike remarked on the disparity between the details of the study, which noted the possibility of threats to drinking water from fracking with toxic chemicals, and the overall conclusion, which stated that fracking was entirely safe. Industry wanted the details changed; activists wanted the conclusion amended to reflect the details.

EPA's Own Contractor Finds Fracking Poses Risks; EPA Dismisses It

Cadmus took over the report in late 2002 from the original contractor. The project faced obstacles from the outset, according to EPA documents and Cadmus staff. The oil and gas industry declined to provide information about the composition of fracking fluids, asserting that they were trade secrets. There wasn't enough time or money for Cadmus to begin monitoring groundwater before, during and after fracking jobs to see if the process affected water quality. With little insight into what was actually pumped into the earth during fracking, Cadmus researchers had to rely on existing literature and discussions with a limited number of experts familiar with the process.

Cadmus sent chapters of its working draft to the EPA starting in mid-2003. The agency immediately questioned the validity of the findings. Against common scientific practice, the EPA urged Cadmus to include an oil industry study that had not been peer-reviewed. When Cadmus staff resisted, the EPA manager asked a Cadmus scientist, "'Can't you say something positive about it?'" the scientist recalled.

The industry study fell by the wayside. But the EPA changed parts of the working draft that suggested fracking for coalbed methane could pose risks to drinking water, according to the documents and Cadmus scientists.

A March 3, 2004, EPA agenda entitled "Hydraulic Fracturing Project Status" listed among the tasks "Soften conclusions and ES [executive summary]."

In drafts of the executive summary, typically a report's most widely read section, the authors referred to potential threats to public health as the reason for the study. "The goal of this Phase I study was to determine if a threat to public health exists as a result of USDW [Underground Sources of Drinking Water] contamination from hydraulic fracturing fluid injection into CBM [coalbed methane] wells, and if it does, whether the threat is great enough to warrant further study," the authors wrote.

The final version of the report omits mention of public health except in the discussion of methodology and in paraphrasing public comments deep into the 463-page study.

EPA documents show Cadmus recommended revisions to reflect complaints by some Virginia residents about possible contamination of their water from fracking. The contractor raised the question of an investigation to see if the complaints were warranted. The final version did not include the changes Cadmus recommended, and EPA did not launch an inquiry into the complaints.

The Cadmus scientists said they realized over time that their findings about risks to underground drinking water diverged from what the EPA wanted. The scientists determined that fracking does pose some risk to drinking water. They concluded that monitoring of fracking activities and more information from industry would be needed to quantify the risk. The EPA decided the study's conclusion should be that fracking did not pose a threat to groundwater and therefore did not require further study or federal oversight.

The Cadmus scientists came to believe that abiding by the EPA's conclusion violated their standards of integrity. "If you say there is no risk associated with hydraulic fracturing, and we see risk, you either didn't do a good job or you're lying," Sham said. "The data and analyses tell us there is risk associated with it, and we were asked to report there is no risk, and we can't say that."

The EPA routinely hires contractors to conduct studies, and the firms' names are generally tucked away in appendices or acknowledgements. Contractors appreciate a mention because if the studies are well-regarded, they serve as a form of marketing. The 2004 coalbed methane study notes the use of a contractor but does not identify Cadmus.

"We had no power over the final report. The only power we had was to take our names off it," said a Cadmus team scientist who declined to be identified because of concerns about job security.

Inside the EPA, some scientists were also troubled by the study. "What I found objectionable was that it was written to have a good P.R. effect on people," said Mario Salazar, an engineer who was an internal reviewer of the report and worked as a technical expert at the EPA's underground injection office. "So that people would read the report and say there was no problem with hydraulic fracturing and water."

After the EPA published the final coalbed methane report in June 2004, Weston Wilson, an environmental engineer in the EPA's Denver office, filed a formal whistleblower complaintabout it. Wilson alleged in his complaint that the study's conclusions were "unsupportable" and based on "limited research."

The EPA's inspector general launched an investigation into Wilson's complaint. But the inquiry was closed after the Republican-controlled Congress passed the Energy Policy Act in 2005, codifying in law the conclusion of the coalbed methane study and exempting fracking from the Safe Drinking Water Act.

Back in Pennsylvania, a Boy's Health Problems Grow

In mid-2012, Chevron Appalachia hydraulically fractured two wells on a hill about 400 feet behind Latkanich's house. They produced gas by winter, and Latkanich got royalty checks that at first were as high as $11,000 a month. He paid off legal bills and his mortgage.

But problems soon cropped up that grew increasingly alarming.

The company carved out a two-acre well pad from the hillside for three large impoundments to hold water from other gas sites that was trucked in to frack the two Latkanich wells. During a hard rain or snow melt, runoff from the well pad flowed down the hill, over the site of Latkanich's well, and into his garage and basement.

Latkanich's drinking water developed a metallic taste over the course of the year. He started to get frequent diarrhea. In early 2013, Ryan, then 3, came out of the tub covered in open sores. Latkanich called the state Department of Environmental Protection (DEP) to test the water. Chevron Appalachia declined to hook up Latkanich's home to the nearby municipal water system and provided him with a large outdoor tank instead. The DEP tests did not show anything wrong with the drinking water, and the company took the tank away.

Still worried even after the 2013 tests, Latkanich began to have bottled water delivered. Because he's on a fixed income, Latkanich and Ryan use it only for drinking. They still cook, brush their teeth, bathe and wash dishes and clothes in well water. Ryan's mother left in August 2013.

In December 2016, Ryan started to soil himself almost daily. He was 6, a chubby precocious redhead with perpetually askew glasses. One day, he soiled himself at school. "Charlie was the smartest kid in the class. He was making fun of me in front of the whole class. He said I stink," Ryan recalled. He doesn't have many friends at school now. "I'll never forget that."

Medical tests found nothing wrong with Ryan. Peer-reviewed science has been mixed so far about the links between fracking and incontinence or gastrointestinal problems among residents who use nearby well water. Latkanich called the state DEP to test his water. He also contacted John Stolz, director of Duquesne University's Center for Environmental Research and Education.

The results have been ambiguous. Unlike most people, Latkanich had an independent lab test his water in 2011 before fracking began, giving him a baseline. In its February 2017 test, the state found increased turbidity, or cloudiness, and the presence of coliform bacteria. DEP officials returned in early November to take more samples. Stolz's test found higher levels of iron, calcium and strontium. The amount of sodium had more than doubled to 510.38 milligrams per liter of water from 238.38 in 2011, before fracking began.

The elevated levels of sodium pose a high risk to Latkanich, who suffers from kidney disease. "The test results prove I can't drink this water," he said.

The ambiguity is typical of water tested near fracking sites. If water quality has worsened, there is seldom a bright line to the fracking. That's partly because under the Halliburton loophole, companies do not reveal everything they inject underground, so labs do not know all the substances they should test for. And in many cases, homeowners enter into settlements with energy companies that prohibit them from revealing what happened.

Chevron Appalachia has not seen Latkanich's 2017 water test results, but a spokeswoman said that past water testing didn't support his claim that fracking affected his water. "Based on a review of the 2011 pre-drill and 2013 post-drill water samples, both Chevron and the DEP concluded that Chevron Appalachia's operations did not affect Mr. Latkanich's water," Veronica Flores-Paniagua said in an email. "We understand that Mr. Latkanich has recently raised the same concerns again regarding his well water. As always, Chevron Appalachia will continue to fully cooperate with the DEP in this matter."

The Legacy of the Now-Disputed EPA Study

When the Energy Policy Act passed, the industry celebrated the exemption of fracking from safe drinking water scrutiny and cited the now-disputed EPA study whenever complaints about pollution arose. In a September 2005 newsletter, the Interstate Oil and Gas Compact Commission said the study had found "no confirmed cases that drinking water wells had been contaminated by hydraulic fracturing fluid injection into coal bed methane wells."

"It was the one big study. You heard it quoted for a decade or more after: that fracking never harmed water," said Greg Dotson, a law professor at the University of Oregon and former lead energy policy staffer for Rep. Henry Waxman of California, the top Democrat on the House Energy and Commerce committee during the Bush era. For members of Congress, "if you wanted to do the right thing, you needed to have data on your side, and this study deprived you of an analytical basis. ... The oil and gas guys always used it. It was instrumental to their winning the debate."

The Trump EPA does not try to hide its intention to roll back rules to help oil and gas. Before taking the reins at EPA, Pruitt built a career based on deep ties to industry. He led a political non-profit funded in part by the petrochemical billionaires Charles and David Koch. He sued the EPA more than a dozen times as Oklahoma attorney general over new pollution standards. As EPA administrator, he has halted or slowed several rules affecting oil and gas.

He has moved to undermine the scientific underpinnings of major rules in part by removing independent academics from the agency's scientific advisory panels that review studies on issues such as fracking. In their place, the Pruitt team has put forth the names of corporate representatives, many drawn from the oil and gas industry, who deny prevailing science on public health hazards such as climate change and ozone.

EPA Issues New Report, But Change is Unlikely

In December 2016, as the Obama administration was about to leave office, the EPA issued a new report, which stated for the first time that fracking in some cases had contaminated drinking water. It identified possible risks to groundwater unless certain safeguards are implemented. Cadmus was the government contractor who helped conduct the study, and this time, its name is repeatedly mentioned in it.

The new study won't change anything on the ground unless Congress acts to repeal the Halliburton loophole, which is unlikely for the present.

Latkanich expects no help from the government. It allowed all he sees around him to happen, he figures. He has a reputation with Chevron as a troublemaker because he monitors and criticizes its practices. Early on, he grew suspicious of the company when he learned from a neighbor that a Chevron contractor had released stormwater runoff into a stream on the other side of his property. The company was cited by the state, but Chevron and state regulators did not tell Latkanich about the violation, he said.

Latkanich would like to stay in his house, which he poured thousands of dollars into because he thought he would grow old in it. Now, his fears about the well water nudge him to go, but he worries he can't find a buyer. "I can't sell the house now: It has foundation issues and pollution. The value of the house has dropped like a rock," he said.

The most likely buyer would be Chevron, and Latkanich is determined to wrest accountability for the damage he believes the company has done. But he can't afford a lawyer to help him negotiate a settlement. One non-profit told him his case was too big and complex for it to handle. He gets a disability check and about $550 monthly now in royalties for his two gas wells, so he doesn't have the money to hire private firms.

"This farm is ruined," he said.

"Forever," said Ryan, who had come into the kitchen from running around outside.

"Buy me out and I'll move somewhere where there isn't fracking," Latkanich said.

"Japan?" Ryan offered. "Because I don't think there's fracking there."

INSTRUMENT OF POWER: HOW FOSSIL FUEL DONORS SHAPED THE ANTI-CLIMATE AGENDA OF A POWERFUL CONGRESSIONAL COMMITTEE

Rep. Lamar Smith has led a strategic attack on climate science using the committee he chairs. Back in Texas, his constituents face the effects of global warming.

By Marianne Lavelle , David Hasemyer
December 5, 2017

FREDERICKSBURG, Texas—It's midway through fall, and cold has yet to settle over the Eckhardt family orchard. So, Diane Eckhardt waits with rising apprehension.

Cold is the switch that triggers the growing sequence that by summer has limbs sagging with ripe, juicy peaches. The reliable chill season in Texas Hill Country allowed Eckhardt's grandfather, Otto, to start the family business here in the 1930s.

But last year, with temperatures the warmest since 1939, Eckhardt's trees produced just 10 percent of their usual yield. And the year before, warm weather reduced production between 60 and 70 percent. Now, Eckhardt worries not only about the next crop, but about the future of a business she hopes will be passed on to her niece and nephews.

"We know climate change is happening," she said.

But while the Eckhardts face that certainty, their congressman sows uncertainty, casting doubt on the consensus science that greenhouse gases are the dominant cause of rising global temperatures, and opposing government action to curb them.

Sixteen-term Republican Lamar Smith has used his power as chairman of the House Science, Space and Technology Committee for the past five years to do battle on behalf of the fossil fuel industry. Embracing the arguments of a small group of climate contrarians, Smith acknowledges that warming is happening but says more research is needed to determine the amount and causes, and whether it does more good than harm.

Smith's critics say he misrepresents facts, cleverly casts doubt on legitimate studies by claiming they are based on "secret data" and uses his subpoena power to help industry battle state and federal regulators and environmental groups. The result is that a panel with vast jurisdiction over all government non-military science, research and development has become an instrument of attack on mainstream climate science.

"Anyone stating what the climate will be in 500 years or even at the end of the century is not credible," Smith said at a hearing he chaired in March. "In the field of climate science, there is legitimate concern that scientists are biased in favor of reaching predetermined conclusions. This invariably leads to alarmist findings that are wrongly reported as facts."

Things weren't always this way. Under both Democratic and moderate Republican leadership, the science committee since the late 1970s had educated lawmakers and the public about the threats posed by rising temperatures caused by human activity and the need for decisive action.

But in a case study of the power of fossil fuel interests to shape government policy, the industry's money and alliances with conservative think tanks and advocacy groups transformed the committee's membership and supported the rise of Smith, son of an old oil and ranching family in South Texas.

Today, "we're in total denial," says the highest-ranking Democrat on the panel, fellow Texan Eddie Bernice Johnson. Instead of looking at the evidence and making policy recommendations on climate—as once happened—the committee is "pretending to be oblivious."

"When you look at the [campaign] contribution list, it becomes very clear" that the forces that oppose regulation are calling the shots, she said.

Smith, 70, has announced he will retire after his term as chairman ends next year. But many believe his legacy will be lasting. The committee under Smith has "contributed to the sort of diminishment of science in public policy," said Andrew Rosenberg, of the Union of Concerned Scientists, an organization that has been one of Smith's targets. "It has reinforced the view that everything must be partisan, and you've got to choose sides. It sends a message that science is partisan, too."

The Science Committee Transformed

The House Science Committee held Congress' first hearings on climate science in 1976, and it resulted in passage of bipartisan legislation to establish a National Climate Program Office. Five years later, Al Gore, then a congressman and committee member, co-chaired another set of hearings on "Carbon Dioxide and Climate: The Greenhouse Effect."

With fidelity to its nonpartisan tradition, the committee championed scientists who found themselves under fire during President George W. Bush's administration. The Republican chairman at the time, Sherwood Boehlert of New York, faced off against Bush officials and GOP colleagues on Capitol Hill, armed, he says now, with what he learned about global warming during 24 years on the committee. "I [witnessed] a parade of some of the most distinguished scientists...from around the world testify that climate change is for real, it has serious and negative consequences, and we damn well better do something about it," Boehlert said.

But by 2013, Republican moderates like Boehlert who accepted climate science were eliminated. Redistricting by both parties—to make red districts redder and blue districts bluer—was partly to blame. But much of the change was due to a systematic effort by conservatives, with significant help from fossil fuel interests that were seeking to stave off policies that might cut into their profits.

Helping the effort was the 2010 Supreme Court decision in *Citizens United v. the Federal Election Commission,* which opened the door to unlimited political spending. The decision empowered hugely wealthy donors like the oil billionaire Koch brothers seeking to shape politics in America. Between 2010 and 2012 alone, individuals and companies in the fossil fuel industry spent an unprecedented $90.5 million to elect friendly Republicans, an increase of 66 percent over the previous two election cycles (compared with $15.7 million they gave to Democrats), according to Center for Responsive Politics data.

In the end, nearly a third of the Republicans who sat on the committee in 2006—seven of 24, including Boehlert—were defeated or retired over the next six years in the wake of primary challenges from the right.

The conservative advocacy group, the Club for Growth, financed many of the challenges. The group has a broad low-tax, less-government agenda, but its victims saw the hand of fossil fuel interests at work.

"My most enduring heresy was saying that climate change was real," said Bob Inglis of South Carolina, who became convinced of climate reality on a science committee trip to Antarctica. He lost in 2010 to Rep. Trey Gowdy, who was endorsed by the Club for Growth. "It had appeared that I had crossed to the other side and had become unfaithful to the tribe."

Wayne Gilchrest of Maryland, a science committee Republican defeated in the 2008 primary, believed the conservative activists had a deeper objection than his stand on climate issues. "The foundation upon which their enterprise was built was to select members of Congress that could be told what to do," Gilchrest said. "They wanted a puppet."

The Club for Growth, which spent between $6 million and $22 million each election cycle between 2004 and 2016, did not have to disclose its donors until it established a Super PAC in 2010. But at least $3 million came into the club from the labyrinth Koch network between 2009 and 2015, according to the tax returns of their various nonprofit advocacy groups. And the club's Super PAC disclosures show that its wealthy donors include some heavily invested in the fossil fuel industry and connected to the Koch brothers, like hedge fund manager Paul Singer of Elliott

Management, and Quantum Energy, a Houston private equity and venture capital firm specializing in oil and gas.

Meanwhile, the fossil fuel industry became the biggest contributor to science committee members, according to Center for Responsive Politics data.

The oil and gas industry has been Smith's biggest contributor, with $764,000 in donations over the course of his career in Congress. Smith points out that the industry's share is just a small portion of his overall contributions— about 5 percent of the $14 million he has raised since the 1989 election cycle.

"I'm supported by a wide variety of individuals and industries, including the energy sector, which employs 400,000 Texans statewide," Smith told InsideClimate News.

After the 2012 election, with GOP moderates on the science committee wiped out, the House Republican leadership chose Smith from among three climate deniers vying for the chairmanship. He had raised four times more money that election cycle than his competitors, F. James Sensenbrenner of Wisconsin and Dana Rohrabacher of California, and had doled out more than $147,000 to help other Republican candidates.

'Secret Science' Charge Returns

Even before Smith took over the committee, he co-signed a letter in December 2012 calling on the Environmental Protection Agency (EPA) to release the "secret data" behind proposed air pollution regulations. On the letter with him was Rep. Andy Harris (R-Md.), a conservative who had unseated Gilchrest and who then chaired the panel's energy subcommittee. "It is likely that the majority of benefits from all federal regulations are grounded in data sets that have never been available to the public," the letter said.

The "secret science" charge was false. At issue was a large, long-term federally funded study published by Harvard researchers in 1993 showing that fine soot pollution, largely from burning fossil fuels, shortened lives. The researchers had obtained personal health information from 22,000 participants on the promise of confidentiality. Fossil fuel interests and advocacy groups they funded rebranded this common health research privacy measure as sinister. In 1997, when the Clinton administration was finalizing the first-ever air quality standards on fine soot, protesters in white lab coats appeared on Capitol Hill holding signs that said, "Harvard, release the data!" The protesters were hired by a group called Citizens for a Sound Economy, which was founded by the Koch brothers and which also received funding from the Exxon Foundation that year.

Since that time, hundreds of studies have affirmed that fine soot causes respiratory and cardiovascular disease and death. And a panel jointly funded by the EPA and the auto industry received access to the Harvard study's raw data for a reanalysis in 2001 that validated the original study.

But 16 years after the white lab coat protests, Smith was reviving the "secret science" charge just as the Obama administration was finalizing a plan to tighten the fine soot regulations, a move vehemently opposed by coal-fired power generators and oil refiners.

Eight months after assuming the chairmanship, Smith slapped the EPA with the first subpoena that the House science committee had issued in 21 years. The subpoena demanded the EPA release data from the Harvard study and a separate American Cancer Society study in sufficient detail "to allow one-to-one mapping of each pollutant and ecological variable to each subject." The EPA worked to obtain the data without personal information, and the agency ultimately released some to Smith, but he was not satisfied. "What is EPA Hiding From the Public?" was the title of a Smith op-ed in the *Wall Street Journal* in June 2014.

Smith drafted legislation called the Secret Science Reform Actto require that the EPA base its regulatory decisions only on scientific data that is publicly available and reproducible. Science advocacy groups say the restriction would curb all regulation to protect public health because health research routinely relies on confidential patient information. It also would rule out regulations based on studies of natural disasters or human-caused events—like spills of oil or fracking wastewater—since they could not be reproduced. And it might put at risk EPA's 2009 finding that carbon pollution was an endangerment to health—the underpinning of all Obama administration action on climate change. The bill is "based on a misunderstanding of how science works," said Rush Holt, the CEO of the American Association for the Advancement of Science, at a hearing earlier this year.

Smith's legislation, which passed the House twice in the Obama era but died in the Senate, was approved by the House again in March with a new name: the HONEST ACT (the Honest and Open New EPA Science Treatment Act of 2017).

"Our goal has been to rely on good science, the facts and reliable data in an effort to discover the truth," Smith told InsideClimate News, when asked to sum up the role he has tried to play. "It is my responsibility to ensure that federal agencies rely on science that has integrity and is free from political influence."

Political Operatives Replace Scientists

Smith's first subpoena came only after a contentious debate and party-line vote in the committee. But his powers were greatly expanded in 2015, when the House leadership allowed him unilateral subpoena power. Smith said it was necessary because, under Obama, "federal agencies often stonewalled the committee's constitutional obligations to conduct oversight." He would issue 25 subpoenas over the next two years, against scientists, regulators, environmental groups and even state attorneys general.

Smith also hired seven staffers from the aggressive House Oversight and Government Reform Committee, which had provided fodder for Republican attacks on the Obama administration with its probes on Benghazi, IRS treatment of conservative groups, solar manufacturer Solyndra's bankruptcy and other issues. "In 2015, it became increasingly apparent that the Obama administration was advancing a one-sided, unconstitutional agenda," Smith told InsideClimate News. "We needed to bring in staff who had strong backgrounds in conducting oversight of government agencies and getting them to answer questions."

The science committee majority staff, which had more than a dozen Ph.D. scientists during the Boehlert era, now was loaded with political operatives like Mark Marin, who had been deputy staff director for Rep. Darrell Issa (R-Calif.) and Joe Brazauskas, who had been a law clerk for the National Mining Association before serving as a counsel on Issa's House Oversight committee.

Warming 'Pause': Smith Investigates NOAA

In June 2015, Smith and his team targeted research published by scientists from the National Oceanic and Atmospheric Administration (NOAA). The paper in the peer-reviewed journal *Science* undercut a key talking point of climate science deniers: It found that there had been no pause in global warming over the past two decades.

In response to Smith's subpoenas and letters, NOAA provided the science committee with its raw data and methodology, and its scientists gave briefings on their research for committee staff. But Smith wanted the researchers' emails.

He said an agency whistleblower had alleged that the paper had been rushed to publication despite the concerns and objections of a number of agency employees. The purported whistleblower, a NOAA scientist named John Bates, later said his complaint had been mischaracterized; he disagreed with how the scientists stored and archived their climate data, but he did not dispute the study's findings or allege data manipulation. "I knew people would misuse this," he told *Science* magazine's website.

Eddie Bernice Johnson fired back at her committee's chairman in a letter: "The only thing you accused NOAA of doing is engaging in climate science—i.e., doing their jobs."

Smith's NOAA investigation provoked an unusual rebuke from the American Association for the Advancement of Scienceand six other major scientific groups, who accused him of chilling scientific inquiry.

Since the NOAA paper's publication, several studies as well as a reanalysis of the original NOAA work have affirmed the finding that there has been no hiatus in global warming.

But Smith continued to accuse the scientists of wrongdoing. In a March hearing, Smith reiterated that the committee heard from whistleblowers that NOAA employees "put their 'thumb on the scale' during the analysis of data." And in an April speech at an annual conference of climate deniers sponsored by the Heartland Institute, Smith said both NOAA and EPA during the Obama administration had been "complicit in furthering a one-sided partisan agenda focused on climate change."

MIT atmospheric scientist Kerry Emanuel believes he gained insight into Smith and his approach after an exchange a couple of years ago. It began with a discussion on a topic on which they agreed—how the U.S. had fallen behind Europe in numerical weather prediction. Emanuel took the opportunity to give Smith a copy of a primer he had written for non-scientists on climate science and risk.

A few days later, Emanuel got a call from Smith, who wanted to talk about the book. "He struck me as a very astute man," Emanuel said. "Clearly he had read the book very thoroughly or had been thoroughly briefed on it.

"He proceeded politely to ask sharp questions. Could this be wrong? Could it be not as bad? A lot of the questions were about uncertainty," Emanuel recalled. At first the scientist felt he was making headway with the congressman, a hope that was quashed the next time he heard Smith publicly dismissing climate science. "In hindsight, I think I was unwittingly a coach, helping him armor himself against reasonable arguments."

"There's nothing stupid about Lamar Smith," said Emanuel. "He's not uniformly anti-science. It's not that he doesn't understand the science. He struck me as a lawyer for the defense, who knows his defendant is guilty, but is bound by law or honor or legal code to defend."

Subpoena Power Unleashed

Smith's tactics to defend fossil fuel energy created legal worries for the EPA, which was battling the industry in court. Officials feared that the constant document requests by Smith would help industry lawyers obtain otherwise confidential material that could be used against the agency in court.

In one instance, Smith acted with a like-minded Republican committee chairman, Jason Chaffetz of House Oversight, to demand internal documents on one of the Obama administration's most contentious regulations, intended to protect thousands of waterways and marshlands. After Chaffetz obtained memos showing an interagency dispute over the rule, he released them. EPA's foes, including the oil industry, which would not normally have had access to the records, sought to introduce them in a court challenge to the new clean water regulation. "Once documents have been disclosed and widely disseminated, an agency has waived any deliberative process or other privilege that may have applied," argued North Dakota's attorney general, the lead litigant.

In letters to Smith, EPA repeatedly raised its concern about the risk of releasing documents in cases involving the Clean Water Act, the regional haze pollution standards, a decision over a controversial copper mine decision in Alaska, and others.

Smith's most dramatic rush to the legal defense of the fossil fuel industry was his unprecedented move last year to issue subpoenas to two state attorneys general and several nongovernmental advocacy groups over the states' climate change fraud investigation of Exxon. He accused the attorneys general of colluding with environmentalists, violating Exxon's free speech rights and chilling private sector science funding.

New York Attorney General Eric Schneiderman, a Democrat, refused to comply with the subpoena, which he said "oversteps the boundaries of federalism, separation of powers, and the committee's own jurisdiction."

Smith disagreed. "The science committee has jurisdiction over all non-military, non-medical research and development," he said in the email to InsideClimate News. "We had an obligation to the scientific community and the American people to find out whether the attorneys general have intentionally intimidated researchers who disagree with them."

Legal experts see another possible outcome of Smith's inquiry. If he were to obtain and make public internal documents like correspondence among the attorneys general—material that Exxon would have trouble obtaining as a litigant—the company could use it in court.

"It sure looks like he is acting on behalf of Exxon," said Maryland Attorney General Brian Frosh, a Democrat who is part of a coalition supporting the investigation by New York and Massachusetts. "It's hard to understand why he thinks that it's an appropriate role for a congressman and Congress to get involved on behalf of an entity like Exxon. It's not like ExxonMobil is incapable of defending itself."

And, indeed, even though Smith did not obtain documents, Exxon cited the mere fact of his inquiry to give weight to its effort last year to derail the probe. Exxon picked up Smith's argument that the attorneys general were appointing "themselves to decide what is valid and what is invalid regarding climate change." For that reason and on multiple constitutional grounds, Exxon lawyers asked a federal judge in Texas to dismiss the investigations.

Although the Texas judge declined to rule, he embraced Exxon and Smith's contention that the New York and Massachusetts investigations were intended to "squelch public discourse by a private company that may not toe the same line as these two attorneys general." The case has been transferred to New York federal court, where a judge has yet to rule.

Back Home in the 21st District

For the most part, Smith hasn't had to address the climate issue back home, even though polls show that a majority of his adult constituents believe human activity is causing global warming and Texas has suffered more severe climate and weather disasters since 1980 than any other state.

Only last year did a Democratic opponent, Tom Wakely, try to make climate change a major campaign issue. Smith won with 57 percent of the vote—the first time he had fallen below the 60 percent mark. Boehlert believes voters don't care as urgently about climate change as they do about jobs, health care and the economy. "People don't think the environment touches them dramatically and personally, but it does," he said.

For Diane Eckhardt, it's hard to ignore what's happening. She kicks up an orange cloud of dust as she walks through her family's orchard at sunset with her 85-year-old father, Donald, and her 9-year-old nephew, Quentin.

Eckhardt, 43, who has a degree in biology, tries to keep politics and peaches separate. But it's clear she disagrees with policymakers like Smith whose distrust of the science imperils her family business and its future.

"We have to pay attention to climate science because the science is there," she said. "We have to extend outside of our ideologies to protect what we have."

HOW BIG OIL LOST CONTROL OF ITS CLIMATE MISINFORMATION MACHINE

One of the longest and most consequential campaigns against science in modern history is becoming more extreme—and turning against its originators.

By Neela Banerjee
December 22, 2017

The Heartland Institute, a conservative think tank, launched a billboard campaign in 2012 to compare believers in global warming to "murderers and madmen" such as the Unabomber, Charles Manson and Osama bin Laden. The backlash was so severe that Heartland pulled the plug within 24 hours, but it still lost major donors and political allies and faced criticism that its fight against climate science was beyond extreme.

Five years later, on June 1, 2017, the group's chief executive, Joseph Bast, was a guest of Donald Trump in the White House Rose Garden as the president announced the withdrawal of the United States from the Paris climate agreement.

"We are winning in the global warming war," Bast declared later in an email to supporters.

Heartland's rebound is striking. Its ascent into the Trump administration's orbit, where it now advises the Environmental Protection Agency on climate change issues, marks the most dramatic success yet in a decades-long crusade, first funded by fossil fuel money, against the mainstream scientific conclusion that human activity is warming the planet and inviting disastrous consequences.

Hundreds of millions of dollars from corporations such as ExxonMobil and wealthy individuals such as the billionaires Charles and David Koch have supported the development of a sprawling network, which includes Heartland and other think tanks, advocacy groups and political operatives. They have cast doubt on consensus science, confused public opinion and forestalled passage of laws and regulations that would address the global environmental crisis. It is one of the largest, longest and most consequential misinformation efforts mounted against mainstream science by an industry. Climate denial, thanks to the network's influence, has become a core message of the Republican Party, now in control of the White House and Congress.

"This didn't come out of nowhere. Trump was taught to say these things on climate by Heartland, the Competitive Enterprise Institute and other think tanks. They maintained this denial space in public policy dialogue," said Kert Davies, director of the Climate Investigations Center, a watchdog group. "And you can definitely credit Exxon and Koch brothers' money for giving the think tanks the megaphone to keep climate science denial in the world."

But now, just like the Republican upstarts that threaten the party establishment, Heartland is taking climate denial farther than many fossil fuel companies can support. While ExxonMobil today publicly accepts the reality of human-caused climate change and the need to address the problem, Heartland argues for the benefits of a warming world. The group is pushing the EPA to overturn its official conclusion—known as the endangerment finding—that excessive carbon dioxide is a danger to human health and welfare. The finding, affirmed by the Supreme Court, is what empowers the agency to regulate carbon dioxide and other greenhouse gases.

This rift was on display at a recent meeting of the American Legislative Exchange Council, a group that influences state governments to adopt conservative priorities. Heartland wanted ALEC to approve a resolution calling on the EPA to withdraw the endangerment finding. But ExxonMobil, once at the forefront of climate denial, was among several corporations and utilities that convinced ALEC to shelve a vote on the resolution.

ExxonMobil had become just another member of "the discredited and anti-energy global warming movement," complained Heartland's president, Tim Huelskamp, a former Republican congressman from Kansas. "They've put their profits and 'green' virtue signaling above sound science."

ExxonMobil is among an early group of donors that slowed or ended its funding of climate denial. But the misinformation apparatus the corporations helped create is now so independent and robust, it no longer needs—or trusts—them.

"Robespierre beheading Danton is pretty apt here," said Jerry Taylor, president of the bipartisan, pro-climate action think tank Niskanen Center, referring to French revolutionaries executing the moderates among them during the Reign of Terror. "There used to be some degree of interest in projecting an image of seriousness, of expertise and evenhandedness on climate, and there isn't anymore."

The new goal is making sure that denial is "part of the ideological catechism of the conservative base," Taylor said. "They are trying to keep the hard Right animated."

Taylor knows this universe from the inside. Once a prominent climate skeptic, he worked at ALEC as staff director for energy and environment issues early in his career. From 1991 to 2014, he was a vice president at the Cato Institute, focusing on energy and climate issues. And, his brother, James Taylor, is a senior environmental fellow at Heartland.

Whatever their differences today, corporations such as ExxonMobil were crucial to getting the denial network up and running.

According to climate watchdogs Greenpeace/ExxonSecrets, ExxonMobil led corporate donations to think tanks, giving nearly $31 million between 1998 and 2014 to 69 groups that spread climate misinformation. The Koch brothers, whose conservative ideology dovetails with their petrochemical business interests, led giving among individual magnates, donating more than $100 million since 1997 to 84 groups.

The Heartland Institute rejects suggestions that it was ever part of any group fostered by corporations. "Heartland has long supported and promoted scientists skeptical of man-caused global warming on principle," Jim Lakely, the organization's spokesman, told InsideClimate News.

Data from Greenpeace/ExxonSecrets shows the organization received $650,000 from ExxonMobil between 1998 and 2006. Heartland points out that it ramped up its work on climate after its Exxon funding ended, and that the only Koch-connected contribution it received in the past 15 years, of $25,000, supported its work on health care issues.

ExxonMobil and Koch Industries did not respond to requests for comment.

As Money Flowed, Strategies Passed Like Baton

Social scientists are still trying to gauge the full breadth of spending on climate denial because of the large number of players involved, the growth of money from secretive sources, and the wide range of public relations tactics fossil fuel companies use to delay action, according to Robert Brulle, professor of sociology at Drexel University.

Brulle mined IRS and other public filings to show that between 2003 and 2010, 91 groups promulgating climate denial received more than a half-billion dollars from 140 foundations. Corporations such as ExxonMobil pared back their funding in that decade; most of the money came from financial vehicles such as Donors' Trust, which shields funders' identities.

As the money flowed through and nurtured the network over the decades, its misinformation strategies passed like a baton to a shifting array of coalitions and initiatives that protected fossil fuel interests in the climate debate. Some groups produced reports that cast doubt on the accumulating evidence of manmade climate change, and others amplified the alternative findings. Think tanks in the network held conferences, sponsored panels, wrote op-eds and letters, and created an echo chamber loud enough to command equal time in the mainstream media.

Still, in 2007, under pressure from its shareholders, ExxonMobil announced in its Corporate Citizenship report that it would stop funding a number of climate denial groups. The following year, the presidential nominees of both major political parties felt safe promising voters they would address the threat of climate change if elected.

Then in 2010 the equilibrium changed. The Supreme Court's decision in *Citizens United vs. FEC* removed caps on corporate and nonprofit political donations and opened the floodgates on campaign spending. Billionaires such as the Kochs moved millions of dollars to support the rise of the Tea Party movement and ultra-conservative candidates who saw climate denial as the bedrock of party orthodoxy. Soon, few Republicans running for federal office would admit to accepting the reality of manmade climate change. After conservative populism put Donald Trump in office, the hard-right's social media ecosystem, empowered by a president eager to tweet, has accelerated the spread of false climate narratives, making them more difficult to counter, much less uproot.

"They decided that if they have to choose between an argument that is solid and serious and one that is dodgy but easily understandable by the base, they'll go with the latter," Taylor said. "Anything that moves the needle for the Fox News or Breitbart reader."

The machinery of climate misinformation has gone global and now runs itself. A false story that appeared last February in the Daily Mail, a British tabloid, got pushed by right-wing outlets and social media deep into the halls of Congress. "Exposed: How world leaders were duped into investing billions over manipulated global warming data," its headline said. The story claimed that American government scientists had manipulated climate research to advance the Paris accord.

Ten days later, Texas Republican Lamar Smith, a fossil fuel champion and chair of the powerful House Committee on Science and Technology, pressed the government to release the data the scientists had allegedly misused. The story of scientific deception was fake news. A British media oversight body forced the Daily Mail to retract the story, but that was six months after it was published.

'Delay and Defeat': A Strategy Evolves

Oil companies began developing strategies to sow doubt about science that could lead to regulation long before global warming became an issue.

Beginning in the 1940s, smog routinely choked Los Angeles, fueling a public health crisis. The city hired Arie Haagen-Smit, a biochemist from the California Institute of Technology, to investigate the cause. He quickly identified oil as the culprit, showing that nitrogen oxide emissions and uncombusted hydrocarbons from car tailpipes and refineries formed smog when exposed to sunlight.

The Smoke and Fumes committee at the American Petroleum Institute (API), the industry's main lobbyist, counter-attacked. It funded scientists who rebutted Haagen-Smit and disparaged him personally. The industry asserted that the science of smog was too uncertain to justify new laws or expensive pollution-control equipment.

By the mid-1950s, industry's own researchers confirmed Haagen-Smit's findings. And the first Clean Air Act, in 1963, soon put industry in the crosshairs of federal regulators. Oil companies began campaigns claiming smog controls would cripple the economy.

API has continued the decades-long fight against stricter smog limits, with new rules delayed along the way by Republican and Democratic administrations alike. The 1970 Clean Air Act, meanwhile, has provided trillions of dollars in health and economic benefits, far exceeding the cost of regulations. One Environmental Protection Agency study calculated that the benefits of the 1990 amendments to the law would amount to $2 trillion a year by 2020.

Louis McCabe, Los Angeles's first smog regulator, summarized in 1949 what would become an enduring industry strategy: "Why have we generally failed in our efforts to control air pollution?" he asked. "We have failed because industry believed that air pollution control cost too much. Smoke and dusts were the wages of a prosperous industrial community...There were 'cooperative' programs with the dual objectives of delay and defeat."

A Corporate Pivot to Uncertainty

As the oil industry bolstered its own research into air pollution in the 1950s, some of its scientists began conducting basic research into carbon dioxide emissions from fossil fuels, following work published by leading academics. In 1968, the industry's main pollution-control consultants warned API that it should pay close attention to carbon dioxide emissions.

By the mid-1970s, Exxon began to take carbon pollution seriously. In July 1977, James Black, a senior scientist at Exxon, told top executives that carbon dioxide emissions from burning fossil fuels would warm the atmosphere and endanger human life. Company leaders knew that if fossil fuel emissions made the planet hotter, politicians would likely take steps to cut pollution. So Exxon launched its own sampling of carbon dioxide in the air and oceans and conducted rigorous climate modeling to better understand how the planet was warming, when temperatures might rise, and the effect on human life. Exxon believed that its own peer-reviewed research would give it a credible voice in policymaking if the government decided to regulate emissions. Other fossil fuel companies followed Exxon's lead.

But the industry's forthright approach started to shift toward denial after landmark events in 1988. NASA's James Hansen, a leading climate expert, raised alarm bells in Congress with his testimony that the warming trend was driven by carbon dioxide emissions. The United Nations that year established the Intergovernmental Panel on Climate Change (IPCC) to provide assessments of evolving climate science. More than 300 scientists and policymakers met at a climate conference in Toronto, declaring, "it is imperative to act now." They called for cutting greenhouse gas emissions 20 percent over two decades. Regulation of carbon pollution went from a distant possibility to an imminent threat to fossil fuel businesses.

In December 1988, API helped sponsor a conference on preparing for climate change. And the oil industry and utility companies began to pivot to a new narrative: There was uncertainty and, thus, no urgency to heed the emerging science. In late 1995, Leonard S. Bernstein, a scientist for Mobil Oil Corp., drafted a primer about climate science for the industry's Global Climate Coalition (GCC). In it, Bernstein wrote: "the potential for human impact on climate is based on well-established scientific fact, and should not be denied." But he later added, "It is still not possible to accurately predict the magnitude (if any), timing or impact of climate change."

As the international community moved in 1997 to curb emissions with the Kyoto Protocol, Exxon's Chairman and CEO Lee Raymond focused on amplifying scientific doubt.

"Let's agree there's a lot we really don't know about how climate will change in the 21st century and beyond," Raymond said in a 1997 speech. "We need to understand the issue better, and fortunately, we have time."

Doubts about climate change were echoed by think tanks that the corporations nurtured with donations starting in the 1990s. Boosted by a grant from Exxon, the Competitive Enterprise Institute organized the Cooler Heads Coalition in 1998, which over time has brought together more than 30 conservative groups into an influential echo chamber of climate denial. The group still exists.

Halting climate action has been a leading priority of this coalition and its allies, not reluctant to promote questionable information if it supports their cause.

In one instance, the Spanish economist Gabriel Calzada Alvarez published a study in 2009 that said Spain's push to develop renewable energy had hurt employment, costing 2.2 jobs for every clean energy job created. The Spanish government and the U.S. National Renewable Energy Laboratory showed that Calzada's methodology was flawed. But this did not stop deniers from embracing his narrative.

Calzada was a fellow at the Center for the New Europe, a free-market think tank funded in part by ExxonMobil and the Koch brothers. Koch-supported advocacy organizations spread Calzada's findings through blog posts and friendly media. Calzada delivered testimony to the U.S. House of Representatives. Three years later, Kenneth Green of the American Enterprise Institute for Public Policy Research (AEI), another conservative think tank, cited the Spanish study in House testimony against federal support of green jobs. AEI received nearly $5 million from the Kochs and ExxonMobil from 1998 to 2012.

Fossil Fuel Fingerprints on Contrarian Research

Think tanks needed support from science, which proved tricky to get, because 97 percent of peer-reviewed articles published about climate change show that it is driven by human activity. So, industry directly funded the work of contrarians within the climate science community.

Among the best known is Willie Soon, a proponent of the theory that solar cycles drive climate change. His notion has been discredited by mainstream science, which determined that the influence of solar fluctuations has been too small to account for the magnitude of modern warming. Yet deniers and politicians cite his papers as evidence that scientists are divided about what causes climate change.

In 2015, it emerged that Soon's research had received hundreds of thousands of dollars in grants from 2003 to 2015 from API, ExxonMobil, the Charles Koch Foundation and the Southern Company, one of the country's largest coal-burning electric utilities. He called his papers "deliverables" in return for the funding.

"I have a big super-duper paper soon to be accepted on how the sun affects the climate system," Soon wrote in a 2009 email to a research specialist with Southern.

Denialist think tanks and politicians convinced many Americans that scientists such as Soon are as numerous as those whose work shows fossil fuel consumption to be the main driver of climate change. A 2016 survey by the Pew Research Center quantified the success of their effort. Fewer than 30 percent of Americans know that the vast majority of climate scientists and the peer-reviewed literature support the conclusion that global warming is manmade.

'The Earth is Greening'

The Heartland Institute's rise to policy prominence marks a break from previous brokers of climate denial, because it promotes a narrative that was once rejected as too extreme and divorced from accepted climate science.

The narrative—that excessive carbon dioxide is beneficial for the Earth—is now backed by some in the EPA and the White House and is deployed as a weapon against the endangerment finding. One of Heartland's policy experts, Kathleen Hartnett White, who has called carbon dioxide "the gas of life," was nominated by the administration to lead the White House Council on Environmental Quality.

The EPA and the White House did not respond to requests for comment.

In his email to supporters, dated Oct. 12, 2017, and leaked to E&E News, Heartland's CEO Bast detailed plans for how to "market" the positive narrative about excessive carbon dioxide. For instance, he suggests convincing the

EPA and the courts that doubling atmospheric CO2 would increase crop yields, and suing companies for not increasing emissions.

Surveys show "we are winning the public opinion battle, since most Americans don't believe global warming is a problem that merits the attention being given to it by the media and politicians," Bast wrote. "The best messages are positive: CO2 increases crop yields, the earth is greening."

Heartland rejects claims that it "denies science." Rather, it asserts that the idea of a scientific consensus that climate change is human-caused is without merit. "The scientists we work with have looked at the data and have concluded climate change is not driven by human activity," Lakely told InsideClimate News. "And the consequences of our uncontrollable warming world are not universally bad, but have a lot of positives, according to the scientists and other experts Heartland communicates and work with."

When the Global Climate Coalition disbanded in early 2002, its members said their work was done because they had succeeded in keeping the United States out of the Kyoto Protocol, the first global climate treaty that paved the way for the Paris Agreement. Now, a dozen years later, climate misinformation is emanating directly from the White House. The administration has orchestrated a rollback of regulatory measures on climate change adopted during the Obama years, and the exit from the Paris accord that President Trump announced with much fanfare from the Rose Garden.

"The basic parameters of the long-term threat posed by climate change were well described and known by 1979," Brulle of Drexel said, referring to a major report on climate change issued by the National Academy of Sciences. "But here we are, coming up on nearly 40 years, and there still is confusion and a lack of willingness to act. So I guess in that sense, the effort to stop climate action has won, as if this is a winning position in any sense of the term."

HOW FEDERAL GIVEAWAYS TO BIG COAL LEAVE RANCHERS AND TAXPAYERS OUT IN THE COLD

Short cuts, subsidies and tax breaks helped create 7,000 jobs in the Powder River Basin. Damage to water, air and land is part of the price borne by the public, too.

By Neela Banerjee, Robert McClure
December 29, 2017

On the morning after the autumn's first snow, L.J. Turner looked out over a creek near his house that reliably watered his family's livestock for more than 70 years. A third-generation Wyoming rancher, Turner remembered hunting rabbits there amid lush marsh grasses and high cottonwood trees when he was a boy in the 1950s.

Then the nation's three largest coal mines began to dig in downhill from his 10,000-acre ranch. To get to the coal, they blasted through and drained the region's aquifers. The marsh grasses vanished. The creek began to recede and eventually ran dry, as did a well Turner dug to feed a livestock watering trough.

As the mines grew and oil and gas wells came, Turner lost not only his ranch's abundant water, but also 6,000 acres he once leased as grazing land—scraped away to reach the coal. Now his cattle herd is half what it was, calves near the coal mines die at alarming rates, and he has had to spend thousands of dollars drilling deeper and deeper wells, eating into an already reduced income.

"You can't really do a whole lot about this as an individual, and there's not a reason or a lot of benefit to sitting there and beating your head against the wall," said Turner, who sued the state unsuccessfully over the loss of water. "It's just, you have to change."

Over the last 40 years, Turner and some of his neighbors have paid a heavy price for the development of energy resources beneath the sagebrush-studded high plains where he lives in the Powder River Basin, a West Virginia-sized swath straddling Montana and Wyoming. The wave of fossil fuel extraction brought jobs and money. It also depleted aquifers that allowed people to live and ranch here for generations, devoured thousands of rural acres, and worsened air quality.

American taxpayers everywhere have paid heavily, too. The federal government owns most of the coal, oil and gas in the ground here. And it has fostered mining and drilling through a host of subsidies, including tax breaks, cheap leases and low royalties that permit fossil fuel corporations to privatize the benefits while socializing many of the costs. Corporations have been able to lease federal coal at $1 a ton or less, use loopholes to halve official royalty rates, and take risks that could push the costs of land reclamation onto taxpayers.

Direct fossil fuel subsidies by the federal government amount to about $10 billion annually in tax breaks and deductions, according to a conservative estimate by Taxpayers for Common Sense, a watchdog group. Oil and gas companies, for instance, can deduct the bulk of a well's drilling costs in the year they are incurred, as opposed to writing off the capital expenditures over many years as other industries do. That deduction cost taxpayers almost $1.5 billion in fiscal year 2015, and "distorts markets by encouraging more investment in the oil and natural gas industry," according to the Treasury Department.

Tallies of direct subsidies don't include the hundreds of millions of dollars in lost tax revenues from undervalued leases and royalty rates. They also don't account for billions taxpayers shell out to clean up after fossil fuel extraction or the resulting damage to human health and the environment from climate change.

After calculating all the effects, the International Monetary Fund estimated that governments worldwide pay more than $5.3 trillion annually to support the burning of fossil fuels. The United States is the second-most prolific fossil fuel subsidizer, behind China, according to the 2017 study.

"We're a century and a half into coal and oil. These are not fledgling industries," said Dan Bucks, former director of the Montana Department of Revenue and a consultant on tax and conservation issues. "We're subsidizing fuels with enormous environmental costs especially climate change, and it can't be justified when we know we can supply our energy needs by other means.

"The subsidies slow the natural transition to a more competitive and environmentally sound energy future," Bucks added.

Industry spends considerable amounts to hold on to this taxpayer support, part of a multi-pronged strategy it deploys across courtrooms, legislatures, news media and in Congress to preserve its bottom line. During the 2016 federal election cycle, fossil fuel companies laid out $354 million in campaign donations and lobbying, according to a report by the climate watchdog group Oil Change International. Over the same period, the industry enjoyed

$29.4 billion in direct and indirect federal and state subsidies, the group estimated.

Many fossil fuel projects would likely have gone ahead even without the tax and royalty breaks. But reducing the number and forcing companies to pay for the impact on air, water, land and the climate would shrink fossil fuel extraction markedly, analysts said.

If coal mining in the Powder River Basin had gone slower or been less widespread, things could have ended up differently for the Turners.

"There are significant external costs that companies don't pay for, and if they had to pay for them, we would have a lot less coal production" in the Powder River Basin, said Mark Squillace, director of the Natural Resources Law Center at the University of Colorado Law School. "And there's no doubt that if you had less coal development there, you would have less aquifer depletion."

The government has long subsidized industries to foster their growth. Renewable energy gets tax breaks—but their cost to taxpayers to date is smaller than for fossil fuels and they're written to sunset in the future. Fossil fuel supports have lasted for decades, with two oil subsidies about a century old.

In 2009, President Barack Obama pledged to phase out fossil fuel subsidies by 2020, citing a need to keep more fossil fuels in the ground and avoid the worst damage from climate change. The Republican-led Congress took no action on Obama's budget requests to end them. In his last year in office, the Interior Department suspended new federal coal leases in order to overhaul the program so it better reflects environmental costs and boosts revenue.

The Trump administration reversed those efforts, part of the president's deregulatory agenda aimed at maximizing production of coal, oil and natural gas while backing away from climate policy. The tax bill signed by Trump this month preserves the subsidies enjoyed by industry. "We've ended the war on beautiful, clean coal," Trump declared recently.

Free Land in Coal Country, But With a Catch

After fighting in World War I, L.J. Turner's father sold the family farm in Missouri and moved a thousand miles to become a rancher in Wyoming. There he got a subsidy of his own. The federal government gave the Turner family land for free under the Homestead Act, passed during the Civil War to encourage westward expansion. When the family was picking acreage, locals advised them, "Be sure you get the big spring there, 'cause it's absolutely foolproof water" for livestock, L.J. Turner said, recalling the stories he was told.

But there was a catch: A few years before the Turners arrived, Congress changed the terms of the Homestead Act so that the federal government would retain ownership of minerals beneath the land it was deeding to settlers. It was a move that would come to benefit today's corporations immensely.

Mining federal coal in the Mountain West began in spurts in the 19th century to feed the railroads, but coal leasing accelerated in the 1970s to meet demands of utilities that needed more fuel to support booming manufacturing. The 1973 Arab oil embargo added urgency.

At the time, environmental activists, empowered by the Federal Lands and Management Policy Act of 1976, were pushing for opening more public lands to uses other than resource extraction and for better planning. Even supporters of resource extraction wanted competitive bids and better royalty rates so that taxpayers would be paid a fair price for the fossil fuels they own.

But "the urgency to mine coal ended up winning out," said Bucks, who served on a U.S. Commerce Department commission in the 1970s that studied the consequences of coal mining in Wyoming and Montana. "Land management was complex, the science evolving. That's a long-term endeavor, while the energy need was immediate."

The rush to mine had its biggest impact on the Powder River Basin. In 1977, the Black Thunder mine opened about 20 miles east of the Turners' home. Five years later came the Antelope Mine, now owned by Cloud Peak Energy, about 15 miles to the southeast. Eventually, between those two projects, Peabody Energy built the largest surface coal mine in the world, the North Antelope Rochelle mine.

By 2002, the Powder River Basin had become the country's largest coal-producing region. Today it produces 40 percent of all the coal burned in the United States and accounts for more than 10 percent of the country's annual greenhouse gas emissions.

The coal boom employed 7,000 people in Wyoming at its employment peak in 2011. By 2016, that number had fallen to fewer than 5,700, as demand for coal slowed considerably under competitive pressure from cheaper natural gas and renewable sources and the weight of federal regulations.

Even with the industry in decline, the government continues to support the coal companies in the Powder River Basin and elsewhere by offering short cuts, subsidies and tax breaks at nearly every step of the mining process.

Taxpayer Coal for Sale at $1 a Ton—or Less

Since coal mining took off there 40 years ago, investigations by federal agencies and independent analysts have repeatedly shown that the leasing process shortchanges taxpayers. The first two coal leases sold in 1982 "were legally suspect and publicly criticized for not receiving fair market value," leading to a major Congressional inquiry, Squillace wrote in a 2013 law journal article.

In 1990, the Interior Department under Pres. George H.W. Bush rescinded the Powder River Basin's status as a federal coal-producing region for reasons that remain unclear. The move loosened coal leasing restrictions. Rather than Interior deciding where, when and how much coal should be mined, those decisions fell to industry.

One enduring effect of this has been to make noncompetitive lease sales the norm. A corporation will pick a tract of coal to expand an existing mine and petition Interior to lease it. These parcels are too small to be standalone mines that would otherwise invite competitors, Squillace said. So when the coal lease is put up for sale, there's usually only a single bidder, the nonpartisan Government Accountability Office (GAO) reported in 2013.

In a survey of 107 leased parcels, 96 of them, or 90 percent, went to a sole bidder, almost always a company seeking to expand an existing mine, the report found.

This lack of competition means rock-bottom rates and sparse returns to taxpayers. Over the years, the federal government has sold leases in the Powder River Basin for about $1.00 per ton of coal or less. The market price of Powder River Basin coal is about $12.00 a ton, according to the Energy Information Administration.

Besides the lack of competition, leases are so cheap because the Interior Department itself sets the initial price for a coal tract at far below market value. The process for calculating a lease's value is confidential, but separate studies by the Interior Department's Inspector General and the GAO determined the process is flawed. It fails to account for growing overseas demand and other market forces, and deprives taxpayers of millions of dollars annually.

The government collects royalties on the sale of the coal. But the way royalties are calculated also benefits coal companies at the expense of taxpayers. The official royalty rate is 8 percent on sales of coal from underground mines, and 12.5 percent for strip-mined coal like in the Powder River Basin. But corporations often convince federal officials to reduce those rates if the companies are facing financial hardship, or if the coal is expensive to mine for technological reasons, according to Pamela Eaton, senior adviser for the energy and climate program at the Wilderness Society, a conservation group.

For instance, in September, the Interior Department signed off on an expansion of Arch Coal's underground West Elk Mine in Colorado—despite warnings about high greenhouse emissions—and cut the royalty to 5 percent because Arch said the coal was especially difficult to mine.

Between 1982 and 2011, taxpayers lost around $28.9 billion from undervalued lease sales and royalty payments, or about $1 billion a year, according to an analysis by the Institute for Energy Economics & Financial Analysis, a think tank working to encourage a transition away from fossil fuels.

Among the Biggest Subsidies, Captive Transactions

The biggest royalty losses to the Treasury occur once coal has been mined and is sold. By law, royalties are assessed on the first sale after the coal has come out of the ground.

Companies have succeeded in minimizing payments to the government by setting up networks of subsidiaries, to which they make the first sale at low prices. The coal then gets sold, and resold at higher and higher prices, until a power plant buys it.

Taxpayers get a royalty payment only on that first, captive transaction.

In 2004, only 4 percent of Wyoming coal was sold through captive transactions between a corporate parent and a subsidiary. By 2012, that figure rose to 42 percent, according to a review of federal data by the Center for American Progress, a liberal think tank. In a 2013 Securities and Exchange Commission filing, Cloud Peak Energy, one the biggest companies in the Powder River Basin, disclosed that if the federal government changed captive transactions in the coal sector, "our profitability and cash flows would be materially adversely affected."

An investigation by Reuters estimated that industry used the loophole created by captive transactions to pay at least $40 million less in royalties in 2011 alone. Even the

Wyoming state auditor recommended stricter federal control over captive transactions among affiliates of a single mining corporation because it had determined that such sales "are highly susceptible to manipulation."

Luke Popovich, spokesman for the National Mining Association, did not answer questions about criticisms of the federal leasing programs, calling the queries "tendentious." He said that coal corporations paid fees "actually above market rates," and that Obama's Interior secretary "concluded no major reform" of the federal coal leasing program was warranted.

In fact, after years of reports critical of federal coal leasing, including by the Interior Department's Inspector General, the Obama administration undertook a sweeping review of the program. In January 2016, the Interior Department under Secretary Sally Jewell paused the sale of new coal leases during this review. In mid-2016, the administration reformed the royalty rule to close the captive transaction loophole, requiring that royalties be assessed on the first sale to an unaffiliated entity.

Just before Trump took office in January 2017, Jewell released the results of the coal leasing review. "Based on the thoughtful input we received through this extensive review, there is a need to modernize the federal coal program," she stated. "We have a responsibility to ensure the public...get a fair return from the sale of America's coal, operate the program efficiently and in a way that meets the needs of our neighbors in coal communities, and minimize the impact coal production has on the planet that our children and grandchildren will inherit."

Jewell's successor, Ryan Zinke, reversed those steps: The results from the review of coal leasing were discarded, the moratorium lifted and the proposed royalty rule rescinded, tilting the system back once more to industry.

Heather Swift, an Interior Department spokeswoman, declined to answer questions about why Zinke overturned Obama-era steps to reform coal leasing. She pointed out that in making its decision the Interior had revived the dormant Royalty Policy Committee to advise Zinke. Nearly all the committee members are from fossil fuel extraction states and industries. There are no local environmentalists, consumer advocates or tax specialists among the representatives.

Cleaning Up After Fossil Fuels

Fossil fuel extraction tears up the land, even when there are no spills or accidents. Forests and prairie get peeled away for mines, well pads, roads and more. Once corporations are finished, they are supposed to restore what they disrupted to a semblance of its previous state. And they are required to post bonds with state and federal authorities, as a form of insurance to pay for restoration—even if they go bankrupt.

The country's largest coal companies often use the option of self-bonding, which allows them to operate without posting any actual cash or collateral, essentially offering their promise that they will pay fully to restore an area once the mine has closed.

Self-bonding rests on the assumption that the corporations are too big and stable to go bankrupt. Yet by 2016, $2.4 billion of the $3.86 billion in outstanding self-bonding obligations nationwide were held by companies that had filed for bankruptcy in recent years, including Alpha Natural Resources, Arch Coal and Peabody. Critics are concerned that taxpayers could be on the hook for reclaiming their old mines if coal sales continue to decline.

Wyoming state regulators proposed rules last month that will tighten self-bonding requirements. Companies will be able to self-bond only up to 70 percent of cleanup costs. Whether they're allowed to self-bond will be based on current credit ratings, rather than older audited financial statements. The rules will no longer allow self-bonding by subsidiary companies, instead forcing the parent mining company to pledge its assets.

Under Jewell, the Interior Department also moved to shore up reclamation, beginning a review of self-bonding to see if it adequately protected taxpayers. Several federal agencies under Obama, including the Justice Department, successfully argued in federal court in 2016 that Alpha Natural Resources should be required to replace self-bonding with outside insurance before exiting bankruptcy.

Now, the Trump Interior Department plans to loosen reclamation insurance standards by permitting routine use of self-bonding once more.

Charlene Murdock, a spokeswoman for Peabody, owner of the North Antelope Rochelle mine near the Turner Ranch, said in an email that the mine "maintains a strong record of environmental stewardship, and our monitoring shows standards to protect air and water quality are being achieved. Peabody works in partnership with neighboring landowners who successfully graze their livestock on the ample forage of reclaimed mine lands."

She added: "In 2016, Peabody's successful land stewardship achieved 1.8 acres of reclamation for every acre disturbed in mining activities. Over the past decade, Peabody has spent $185 million to restore approximately 48,000 acres of land."

Even when coal companies reclaim the land they've stripped away, it's unclear whether the water will return to ranchers such as the Turners who watched it dry up.

"We don't know yet as there has not been a surface mine that has achieved full and final reclamation, including hydrologic reclamation, in Wyoming," said Jeremy Nichols, climate and energy program director at Wild Earth Guardians, an environmental law nonprofit. "We question whether full hydrologic reclamation can be accomplished in the arid Powder River Basin. The coal companies obviously claim they can. We haven't seen anything empirical that suggests this is the case."

For the Turners, Digging Deeper to Find Water

For L.J. and Karen Turner, the effects of fossil fuel extraction have rippled through their property like waves of an earthquake. "It's negatively affected our quality of life—air quality, water quality and our bottom line," Karen Turner said.

They used to graze cattle on large stretches of federal land. After coal mining began, L.J. Turner lost access to 6,000 acres his family had leased for generations near the North Antelope Rochelle mine. It costs about $2 per cow per month to run cattle on public lands. The privately owned pastures the Turners now must use charge about $25 per cow per month, one of the reasons the couple has reduced their herd from 400 head to 200. The Turners have noticed that calves on the remaining piece of grazing land they lease beside a mine are dying at higher rates than those farther away.

When the Turners married in 1969, the water was only about four feet below ground level near their house, Karen recalled. After the first wave of mining in the 1980s, even more water was drained in the 1990s for the extraction of shallow methane, or natural gas known as coalbed methane.

Government scientists once predicted that the drawdown of water because of mining would extend no more than five miles from coal mines. But billions of gallons have been lost across the basin. For many years, state geologists have documented falling underground water levels—up to 582 feet at one well last year, which is deeper than the Washington Monument is tall. Across the wells checked, the average decline was 82 feet last year.

In Gillette, the largest town in the Powder River Basin, population growth fueled by mining made great demands on the region's limited water supplies. In 2007, the city scrambled to come up with the more than $200 million it cost to run water lines about 45 miles to the next county. State taxpayers funded most of the project, which is not yet complete.

The Turners have paid many thousands of dollars to have their wells redrilled to greater depths and have had to install thousands of yards of pipe to supply the family and their livestock.

But the Turners themselves find the deeper well water undrinkable. "It smells like rotten eggs," Karen Turner said. "It doesn't taste too bad if you hold your nose, which is why we buy bottled water." The well water also sometimes forms a gelatinous sludge in toilet tanks, she said.

Environmental rules do little to protect local residents from the loss of their water due to fossil fuel extraction. In 2007, with a neighboring ranching family, the Turners sued the state, saying officials had failed to settle "questions regarding management of Wyoming's most valuable and finite resource: water."

"This is a matter of great public importance, impacting the social and economic realities of the present-day organization of Wyoming's society," the suit said.

The Wyoming Supreme Court rejected the argument, ruling that even though the state conceded that "the administration of water is unquestionably a matter of great importance in Wyoming's arid environment," the ranchers had failed to pursue administrative remedies and "it is not the function of the judicial branch to pass judgment on the general performance of other branches of government."

Standing next to the now-trickling creek where he hunted rabbits and his daughter was married in the 1990s, Turner said he knows he cannot prove in the courts that the coal mines were responsible for the loss of water.

But he knows this: "There was willows and rose bushes and all this kind of brush along here. We'd have beaver, mink, different things like that, sometimes some pheasants. Now it's just, it ain't there no more.

"And whenever we had the water in the creek, frogs were absolutely thick, and now I don't know that we have a frog on the place."

ABOUT US

InsideClimate News is a Pulitzer Prize-winning non-profit, non-partisan news organization that provides essential reporting and analysis on climate, energy and the environment for the public and decision makers. We serve as watchdogs of government, industry and advocacy groups and hold them accountable for their policies and actions. Already one of the largest environmental newsrooms in the country, ICN is committed to establishing a permanent national reporting network, to training the next generation of journalists, and to strengthening the practice of environmental journalism.

We have grown from a founding staff of two to 15 full-time professional journalists and a growing network of contributors.

Climate and energy are defining issues of our time, yet most media outlets are financially unable to devote sufficient resources to environmental and investigative reporting. Our goal is to meet the growing need for coverage that can better inform members of the public in our democracy.

To help keep environmental journalism alive, donate to InsideClimate News by visiting:

https://insideclimatenews.org/about/membership

www.ingramcontent.com/pod-product-compliance
Lightning Source LLC
Chambersburg PA
CBHW072014230526
45468CB00021B/1465